CAMBRIDGE LIBRARY COLLECTION

Books of enduring scholarly value

Physical Sciences

From ancient times, humans have tried to understand the workings of the world around them. The roots of modern physical science go back to the very earliest mechanical devices such as levers and rollers, the mixing of paints and dyes, and the importance of the heavenly bodies in early religious observance and navigation. The physical sciences as we know them today began to emerge as independent academic subjects during the early modern period, in the work of Newton and other 'natural philosophers', and numerous sub-disciplines developed during the centuries that followed. This part of the Cambridge Library Collection is devoted to landmark publications in this area which will be of interest to historians of science concerned with individual scientists, particular discoveries, and advances in scientific method, or with the establishment and development of scientific institutions around the world.

A Selection of Photographs of Stars, Star-Clusters and Nebulae

A geologist and fellow of the Royal Astronomical Society, Isaac Roberts (1829–1904) made significant contributions to the photography of star-clusters and nebulae. By championing reflecting rather than refracting telescopes, Roberts was able to perceive previously unnoticed star-clusters, and was the first to identify the spiral shape of the Great Andromeda Nebula. Roberts' use of a telescope for photographing stars, and a long exposure time, provided greater definition of stellar phenomena than previously used hand-drawings. Although Roberts' conclusions about the nature of the nebulae he photographed were not always correct, the book is significant for the possibilities it suggests for nebular photography. Published in London in 1893 and 1899, the two-volume *Photographs of Stars*, represents the summation of his work with his assistant W.S. Franks at his observatory in Crowborough, Sussex. Volume 1 contains 51 collotype plates of stars, and descriptions of his instruments and methods.

Cambridge University Press has long been a pioneer in the reissuing of out-of-print titles from its own backlist, producing digital reprints of books that are still sought after by scholars and students but could not be reprinted economically using traditional technology. The Cambridge Library Collection extends this activity to a wider range of books which are still of importance to researchers and professionals, either for the source material they contain, or as landmarks in the history of their academic discipline.

Drawing from the world-renowned collections in the Cambridge University Library, and guided by the advice of experts in each subject area, Cambridge University Press is using state-of-the-art scanning machines in its own Printing House to capture the content of each book selected for inclusion. The files are processed to give a consistently clear, crisp image, and the books finished to the high quality standard for which the Press is recognised around the world. The latest print-on-demand technology ensures that the books will remain available indefinitely, and that orders for single or multiple copies can quickly be supplied.

The Cambridge Library Collection will bring back to life books of enduring scholarly value (including out-of-copyright works originally issued by other publishers) across a wide range of disciplines in the humanities and social sciences and in science and technology.

A Selection of
Photographs of
Stars, Star-Clusters
and Nebulae

*Together with Information Concerning the
Instruments and the Methods Employed in the
Pursuit of Celestial Photography*

VOLUME 1

ISAAC ROBERTS

CAMBRIDGE
UNIVERSITY PRESS

CAMBRIDGE UNIVERSITY PRESS

Cambridge, New York, Melbourne, Madrid, Cape Town, Singapore,
São Paolo, Delhi, Dubai, Tokyo

Published in the United States of America by Cambridge University Press, New York

www.cambridge.org
Information on this title: www.cambridge.org/9781108015226

© in this compilation Cambridge University Press 2010

This edition first published 1893
This digitally printed version 2010

ISBN 978-1-108-01522-6 Paperback

SELECTION OF PHOTOGRAPHS

OF

STARS, STAR-CLUSTERS AND NEBULÆ,

TOGETHER WITH

INFORMATION CONCERNING THE INSTRUMENTS AND THE METHODS EMPLOYED
IN THE PURSUIT OF CELESTIAL PHOTOGRAPHY.

BY

ISAAC ROBERTS, D.Sc., F.R.S.,

Fellow of the Royal Astronomical Society; Fellow of the Geological Society; Past President of the Liverpool
Astronomical Society; and of the Liverpool Geological Society.

London:

THE UNIVERSAL PRESS, 326, HIGH HOLBORN, W.C.

THE COLLOTYPE PLATES BY THE DIRECT PHOTO-ENGRAVING CO., BARNSBURY, N.

PREFACE.

———

IT has been my aim, in publishing the photographs and descriptive matter introduced in the following pages, to place data in the hands of Astronomers for the study of astronomical phenomena, which have been obtained by the aid of mechanical, manipulative and chemical processes of the highest order at present attainable, and that such data should be, as regards the photographs, free from all personal errors.

The photographs portray portions of the Starry Heavens in a form at all times available for study and identically as they appear to an observer aided by a powerful telescope and clear sky for observing.

Absent are the atmospheric tremors, the cold observatory, the interrupting clouds, the straining of the eyes, the numbing of the limbs, the errors in recording observations, and the many hardships incurred by our predecessors, of glorious memory, in their attempts to see and fathom the ILLIMITABLE BEYOND.

I commend the observations and the photographs herein to Astronomers and Students of the New Astronomy.

ISAAC ROBERTS.

STARFIELD,
CROWBOROUGH HILL,
SUSSEX,
December, 1893.

Plate 1.

ISAAC ROBERTS' OBSERVATORY.

PHOTOGRAPH

OF

ISAAC ROBERTS' OBSERVATORY

ON THE SUMMIT OF CROWBOROUGH HILL, SUSSEX,

IN

Latitude N. 51° 3′ 7″. Longitude E. 0h. 0m. 37s. Altitude, 780 feet 7 inches
above the mean level of the Sea.

Plate 2.

ISAAC ROBERTS' TELESCOPES.

PHOTOGRAPH

OF

ISAAC ROBERTS' TELESCOPES.

A Reflector of twenty inches aperture and ninety-eight inches focal length for photographic purposes, and a Refractor of seven inches aperture for eye observations.

The two telescopes are mounted on a Declination axis which is common to both, and one clock, with friction governor, moves them together in Right Ascension, but in Declination they can be moved independently.

CONTENTS.

———

LIST OF THE PLATES.

[12]

LIST OF THE PLATES—*continued.*

A LIST OF THE ABBREVIATIONS ADOPTED IN THIS WORK.

N.G.C.—A New General Catalogue of Nebulæ and Clusters of Stars, by Dr J. L. E. Dreyer. Published in the Memoirs of the Royal Astronomical Society, Vol. XLIX., Part I.

G.C.—Sir J. F. W. Herschel's Catalogue of Nebulæ and Clusters of Stars. Published in the Philosophical Transactions of the Royal Society for the year 1864, Vol. CLIV., Part I.

h.—Sir J. F. W. Herschel's Observations of Nebulæ and Clusters of Stars. Published in the Philosophical Transactions of the Royal Society for the year 1833.

D.M.—Durchmusterung by Argelander. Bonn, 1859.

N.—North. S.—South.

F.S.—Fiducial Star ; marked (·), (··), (·.·), (·.·.).

THE NEGATIVES,

From which the Photographs contained in this Publication have been enlarged, measure 10 centimetres square, and one Equatorial Degree upon them measures 44·2 millimetres.

All the Plates have been enlarged from the negatives by photographic methods, to the scales given in the letter-press referring to each subject.

ARRANGEMENT OF THE PHOTOGRAPHS.

The Plates are arranged in the order of *Right Ascension*, and facing each is the descriptive matter concerning the chief features of the objects, together with particulars which will make them available for scientific investigations.

The Plates are placed so as to represent the objects as they would be seen in an inverting telescope, and therefore the edge next to the printed headings is the *south*, and the lower edge the *north*. The right hand side is the *following*, and the left the *preceding*.

The scales of the Plates are given in the letter-press annexed to each.

EPOCH OF THE FIDUCIAL STARS—THE YEAR A.D. 1900.

CERTAIN stars on each of the Plates are marked with dots, numbering from one dot to four or five, thus (::), and the co-ordinates of these stars are given for the epoch A.D. 1900.

The computations have been made from the data found in such of the existing catalogues of stars, as could be made available for the purpose; but the degree of accuracy cannot be absolute, for reasons well known, which arise from proper motions, precession, and errors of observation, and the positions of some of the stars available have not been determined with the highest degree of precision. But the indicated stars will serve the purpose of finding on the Plates the positions of the other stars with considerable accuracy by scale measurements.

TABLE

OF

CORRECTIONS TO BE APPLIED TO THE SCALES OF THE PHOTO-PLATES.

The Collotype photo-processes by which the copying and printing of the plates in this work have been executed, have introduced certain errors in the scales, due to the expansion and contraction of the sensitized films, and to the paper used. Consequently, corrections have to be made to the scales, that they may be applicable, with approximate or practical accuracy, in the measurements of the co-ordinates of the stars. The following is a table of the corrections required :—

Plate 3.—1 millimetre equals 30·10 seconds of arc instead of 30 seconds.

„	4.—1	„	24·16	„	„	24	„
„	5.—1	„	24·32	„	„	24	„
„	7.—1	„	24·33	„	„	24	„
„	8.—1	„	24·58	„	„	24	„
„	9.—1	„	24·27	„	„	24	„
„	11.—1	„	24·28	„	„	24	„
„	13.—1	„	24·11	„	„	24	„
„	14.—1	„	24·22	„	„	24	„
„	15.—1	„	23·79	„	„	24	„
„	16.—1	„	24·22	„	„	24	„
„	17.—1	„	24·24	„	„	24	·,
„	20.—1	„	24·10	„	„	24	„
„	21.—1	„	24·25	„	„	24	·,
„	22.—1	„	24·11	„	„	24	·,
„	23.—1	„	24·35	„	„	24	„
„	24.—1	„	24·49	··	„	24	„
„	26.—1	„	24·42	„	„	24	„
„	27.—1	„	24·27	„	„	24	„
„	32.—1	„	24·32	„	„	24	„
„	38.—1	„	24·11	„	··	24	„
„	39.—1	„	24·55	„	„	24	„
„	40.—1	„	15·50	„	„	15	„
„	42.—1	„	24·10	„	„	24	„
„	43.—1	„	32·34	„	„	31·7	„
„	44.—1	„	15·26	„	··	15	„
„	45.—1	„	24·44	„	„	24	„
„	48.—1	„	24·15	„	··	24	„
„	49.—1	„	24·51	„	„	24	„
„	51.—1	„	24·46	„	„	24	„
„	52.—1	„	24·46	„	„	24	„
„	53.—1	„	15·45	„	„	15	„

A TABLE

For converting the measured *Right Ascensions* of the Stars shown on the Photographs (which are to the scale of 1 millimetre to 24 seconds of arc) into intervals of time at each Degree in Declination between the Equator and the Pole.

DECLINATION.	1 MILLIMETRE =	DECLINATION.	1 MILLIMETRE =	DECLINATION.	1 MILLIMETRE =
	s.		s.		s.
0°	1·60 in R.A.	30°	1·85 in R.A.	60°	3·19 in R.A.
1	1·60 ,,	31	1·86 ,,	61	3·30 ,,
2	1·60 ,,	32	1·88 ,,	62	3·41 ,,
3	1·60 ,,	33	1·90 ,,	63	3·53 ,,
4	1·60 ,,	34	1·92 ,,	64	3·65 ,,
5	1·60 ,,	35	1·95 ,,	65	3·79 ,,
6	1·60 ,,	36	1·98 ,,	66	3·94 ,,
7	1·61 ,,	37	2·00 ,,	67	4·08 ,,
8	1·61 ,,	38	2·03 ,,	68	4·26 ,,
9	1·62 ,,	39	2·05 ,,	69	4·46 ,,
10	1·62 ,,	40	2·08 ,,	70	4·67 ,,
11	1·63 ,,	41	2·11 ,,	71	4·90 ,,
12	1·63 ,,	42	2·15 ,,	72	5·16 ,,
13	1·64 ,,	43	2·18 ,,	73	5·46 ,,
14	1·64 ,,	44	2·22 ,,	74	5·84 ,,
15	1·65 ,,	45	2·26 ,,	75	6·21 ,,
16	1·66 ,,	46	2·30 ,,	76	6·64 ,,
17	1·67 ,,	47	2·34 ,,	77	7·01 ,,
18	1·68 ,,	48	2·39 ,,	78	7·68 ,,
19	1·69 ,,	49	2·43 ,,	79	8·34 ,,
20	1·70 ,,	50	2·48 ,,	80	9·27 ,,
21	1·71 ,,	51	2·54 ,,	81	10·25 ,,
22	1·72 ,	52	2·59 ,,	82	11·45 ,,
23	1·74 ,,	53	2·65 ,,	83	13·27 ,,
24	1·75 ,,	54	2·71 ,,	84	15·37 ,,
25	1·76 ,,	55	2·78 ,,	85	18·12 ,,
26	1·77 ,,	56	2·85 ,,	86	23·36 ,,
27	1·79 ,,	57	2·93 ,,	87	30·74 ,,
28	1·81 ,,	58	3·01 ,,	88	48·66 ,,
29	1·83 ,,	59	3·10 ,,	89	97·33 ,,

ILLUSTRATIONS

OF the method for determining by measurements from the Fiducial Stars the approximate Right Ascension and Declination of any Star on the Plates.

Required the R.A. and Dec. of a Star (x), with a bright companion 44·3 millimetres *south following* F.S. (·), and 108·4 millimetres *north preceding* F.S. (··) on plate 4.

The measurements are made by the direct application of a millimetre scale, or by extending the points of a dividing compass from the centre of the Fiducial Stars to the centre of the selected Star (x), and then reading off the distance on the scale, and are here given for the purpose of identifying the star referred to on the Plate, which will be at the intersection of arcs with radii 44·3 and 108·4 millimetres, but they are not required in the determination of its co-ordinates.

PLACES OF THE FIDUCIAL STARS ON PLATE 4.

F.S. (·) R.A. 0h. 47m. 7·5s. ... Dec. N. 60° 33·9′ ... Magnitude 5·0

,, (··) ,, 0h. 50m. 45·6s. ... ,, ,, 59° 49·3′ ... ,, 6·3

,, (·⋮) ,, 0h. 51m. 16·1s. ... ,, ,, 60° 53·1′ ... ,, 7·0

,, (∷) ,, 0h. 53m. 57·5s. ... ,, ,, 59° 58·3′ ... ,, 7·2

TO FIND THE RIGHT ASCENSION.

It will be observed that the R.A. of F.S. (··)= 0h. 50m. 45·6s.

,, ,, ,, ,, ,, ( )=0h. 47m. 7·5s.

0h. 3m. 38·1s.=

218·1 seconds (the difference between them in R.A.), and by referring to the Table Dec. 60°, which is, omitting fractions, the Declination of the F.S. (·) nearest to the Star (x), it will be seen that one millimetre is equal to 3·2 seconds of time, therefore,

$$\frac{218·1}{3·2}=68·15 \text{ millimetres.}$$

Take this distance between the points of the dividers; place one point on the centre of F.S. (··), and extend the other point towards the *preceding* side; at the same time place the rule with its edge in *north* and *south* direction to bi-section with both F.S. (·) and the disengaged point of the dividers.

The edge of the rule is then on the meridian (0h. 47m. 7·5s.) of the F.S. (·), and whilst the rule is firmly held in this position, measure the distance of Star (x) from it,

c 2

which will be found to be 12·2 millimetres. Then 12·2 × 3·2=39 seconds of time, being the distance in R.A. of the Star (x) from F.S. (•), and 0h. 47m. 7·5s.+39s = 0h. 47m. 46·5s.=the required R.A. of the Star (x).

TO FIND THE DECLINATION.

The scale of the Plate is one millimetre to 24 seconds of arc, and this applies to most of the Plates.

$$\text{Dec. F.S. } (\because)=60° \ 53·1'$$
$$\text{,, \quad ,, } (\cdot)=60° \ 33·9'$$
$$\overline{19 \ 2}=\frac{1152''}{24''}=48 \text{ millimetres}—$$

the difference in Declination between the two F.S. Take this distance between the points of the dividers and place one point on the centre of F.S. (∵), and extend the other point towards the *south;* at the same time place the rule with its edge in the direction of a parallel of Declination and bi-secting both F.S. (·) and the disengaged point of the dividers.

The edge of the rule is then on the parallel of Declination of F.S. (·), and whilst the rule is firmly held in this position, measure the distance of Star (x) from it, which will be found to be 41·8 millimetres. Then 41·8 × 24″=16·7 minutes of arc.

Therefore Dec. F.S. (·)=60° 33·9'—16·7'=0° 17·2', the required Declination of the Star (x), and its position is in R.A. 0h. 47m. 46·5s.; Dec. N. 60° 17·2'. Refraction has not been taken into account in these illustrations, but, if required, the correction could be applied in the usual way.

Methods for determining mathematically the co-ordinates of stars on a photograph have been explained by Sir R. S. Ball and Dr. Rambaut in the *Trans. of the Irish Academy*, vol. 30, part 4, January, 1893; and by Mr. H. H. Turner in *The Observatory*, vol. 16, p. 373, November, 1893.

INTRODUCTION.

THE photographic work which will be described and illustrated in the pages following was preceded in the year 1883 by investigations relative to the suitability of ordinary photographic lenses for the delineation of celestial objects. The method adopted was to attach two cameras to a seven-inch equatorially-mounted refractor. One of the cameras had a portrait lens of two inches aperture, of known good quality, affixed to it, and to the other camera were in succession affixed a selection of other lenses, for comparison with it, having apertures varying up to five inches respectively in diameter. In these cameras were placed photographic dry plates, which were exposed in pairs simultaneously to certain parts of the sky during equal intervals of time, and, in order to secure as nearly as possible equality in all the conditions of comparison, the plates were cut in halves, one-half being placed in each camera.*

The result of these investigations, and of counsel afterwards taken with Sir Howard Grubb, was an order given to him to make and completely mount for me (according to the plans evolved between us) a silver-on-glass reflector of twenty inches clear aperture and one hundred inches focal length. The photographs were to be taken directly in the primary focus of the mirror inside the tube, so as to avoid loss of light by reflection of the stellar images from a plane, or prism, to the outside of the tube in the way usual with reflectors.

The instrument was finished and mounted by the month of April, 1885, when photographic work was immediately commenced; but it was soon discovered that the instrument had several defects, though the works of the optician and mechanician were performed with all the care and skill consistent with a high standard in those days, but of which the results were not sufficiently perfect to meet the requirements of the celestial photographer.

---

* " Proceedings of the Liverpool Astronomical Society," February, 1884.

Errors of eight or ten seconds of arc, by the displacement of a star image, as seen in the ocular during an interval of two or three minutes of time, were of little consequence, and could be easily corrected in a telescope constructed to be used for eye observations. But a displacement equal to three seconds of arc, occurring once in four hours, is not admissible in stellar photography, because that amount of error would distort, during exposure, the images of faint stars, so that they would not appear as circular spots on the developed photograph.

After the consumption of much time, and of other requisites, in localizing the causes of one error after another and attempting to apply remedies, it was found that the greater part of the mounting would have to be reconstructed before the causes of the errors could be eliminated, and also that much greater perfection in the works of the driving clock and sector would be essential before satisfactory photographs, with long exposures of the plates, could be produced. Moreover, a method of controlling the clock by a pendulum would be desirable.

These changes were duly made and the instrument remounted, when again, upon trial, the star images were found to be elongated on the photo-film during long exposures of the plates.

It was at this stage, in the year 1886, that some of the early photographs, showing vast extensions of nebulosity in the Pleiades and in Orion, were presented to the Royal Astronomical Society, and though the images of the stars were elongated, and parts of the nebulous regions in Orion over exposed, and consequently too dense to print on the paper enlargements made from the negatives, yet the extensions and fine details of the faint nebulosity that were shown, particularly on the negatives, carried our knowledge of the structure of the Orion nebula and of nebulæ in the Pleiades far beyond the limits previously attained.

One of the objects I had in view, when deciding to pursue celestial photography, was the charting of the stars in the sky, between the north pole and the equator, on a scale of about twice that adopted by Argelander in the *Durchmusterung*. On May 1st, 1885, the charting was commenced,* and continued till some time after the late Admiral Mouchez, director of the Paris Observatory, in a letter addressed to the President of the Royal Astronomical Society, and read at the meeting held in November, 1885, proposed

---

* " Monthly Notices," R.A.S., Vol. 46, pp. 99-103.

that the charting should be done by the astronomers of all nations, on a uniform plan to be agreed upon between them. With this object in view, a number of astronomers, from all the principal nations of the world, were invited by Admiral Mouchez, to meet in Paris in the year 1887, to discuss and arrange the scheme for charting the stars. How admirably this was arranged and brought to maturity, under the guidance of Admiral Mouchez and Dr. Gill, is now matter of history, and the work which I had previously begun was consequently superseded. But other fields for photographic work are open with ample room for all who choose to labour in them. The field in which I have worked will be recognized by the illustrations and descriptions found in the pages following.

UTILITY OF THE PHOTOGRAPHIC CHARTS.

I have been asked the question, What would be the utility of the charts when they were published, for if any doubts should arise about an appearance upon the chart-prints, they could not be settled without a reference to the original negatives, and these would be less reliable than the sky itself? What force there may be in the objection is, I think, removed by the suggestion which accompanied the question; for, as long as the negatives remain in existence, reference could be made to them. But such an objection could, in its widest application, only affect an occasional star or other object on a chart; whilst hundreds or thousands of stars that may be shown would, within definite small limits of error, be in their true relative positions to each other, and, since the scales of the charts are large, fairly accurate measurements can be made upon them, by use of scale and compasses.

I have examined many negatives (which were taken in duplicate at different intervals of time) by superposing the plates, and bringing the star images into coincidence on corresponding pairs. In this simple way, a difference of three seconds of arc in the coincidence of any star, amongst several hundreds upon the plates, can be easily and quickly detected by eye observations, without having recourse to measurements.

A similar mode of procedure can also be adopted in the examination of the charts on paper, when they have been made to a uniform scale, from negatives taken of the same region of the sky; for, if two charts are superposed, with the stars placed in coincidence, and they are looked at towards a strong light, the non-coincidence of any star, if it exceeds five seconds of arc on the duplicate charts, can be detected.

If, however, measurements to a second of arc are required to be made, then the original negatives must be used for the purpose. But I apprehend that some of the uses to which the printed charts will be applied are the following :—

       1st. The detection of changes in the structure of nebulæ.

       2nd. The detection, on a large scale, of movements amongst the stars.

       3rd. Determinations of variations in stellar magnitudes.

       4th. Relative distribution of the stars in space.

       5th. Detection of new stars and disappearance of others.

The charts will be greatly increased in importance, as the years elapse, from the dates of the photographs. We have evidence of the importance of comparing photo-charts with other charts which have been prepared by hand-work some years previously, and will illustrate this with the following examples :—

In the year 1863, D'Arrest* prepared a chart of the stars which he could see between R.A. 0h. 15m. and R.A. 0h. 19m., with Declination between 63° 0′ and 64° 0′ North. The stars thus charted number 212 (the photograph shows over 400), and about the centre of the space indicated, Tycho Brahe, in the years 1572 to 1574, observed variable stars.

In January, 1890, I took a photograph of this region and enlarged it to D'Arrest's scale, when it was evident, by mere inspection of the configurations of the stars on the two charts, that considerable changes had taken place among the stars on this small area of the sky (comprised within one degree of declination and half a degree in right ascension) during the interval of twenty-six years between the epochs of the charts.

But the question arises, are the changes due, in part or in the whole, to errors made in the charting by hand-work? The photograph is free from any suspicion of error; but no work, however carefully it may have been done by hand and by computation, can be above suspicion.

The statements here made apply also to other charts which have been compared with photo-charts, showing changes which may, or may not, be due to errors in charting. Hence we conclude that in future large contributions to astronomical knowledge must chiefly be looked for from the examination and correlation of photo-charts which have been taken at intervals of a few years, and comparison of the stars and nebulæ upon them. When any changes are observed to have taken place, they can, with confidence, be accepted as realities, and dealt with accordingly.

The charts will also serve as indexes, to point out where changes have taken place among the stars, which can afterwards be subjected to rigorous investigation by aid of the meridian circles and equatorials.

---

* " Trans. Royal Danish Society," January, 1864.

REFRACTORS AND REFLECTORS AS PHOTO-INSTRUMENTS.

The refractor is an instrument which can be brought into and maintained in adjustment much more easily than the reflector, and is, for this reason, better adapted for the charting of the stars; but for the delineation of faint stars and nebulosity, the evidence is so far in favour of the reflector, provided it is brought under the necessary control.

We should expect a higher photo-effect by the reflector than by the refractor, for the reason that a good film of silver on glass reflects more than ninety per cent. of the incident light which is made to impinge without secondary reflection upon the sensitized film, whereas the four surfaces of the photographic lenses in the refractor, together with the absorption of light in its passage through the glass, cannot give results so high, and there is also a further loss in the light of red stars on account of the imperfections which are unavoidable in the chromatic corrections.

REQUIREMENTS AND ADJUSTMENTS OF A REFLECTOR FOR CELESTIAL PHOTOGRAPHY.

If an impression should exist in the mind of anyone that it is a simple and easy operation to take fairly perfect photographs of celestial objects, which will require an exposure of the plates during four hours and upwards in a large instrument, a few trials would in practice carry conviction that the assumption is not well founded. And since there are at present no published treatises as aids in overcoming the difficulties of the work, a brief account of my own experience may be useful to others. The experience was by no means easily or cheaply gained.

Before deciding upon the telescope to be adopted for celestial photography, it would be wise for the intending photo-astronomer to decide upon the field in which he intends chiefly to work, and then select an instrument which will be suitable for the purpose.

The photo-instrument that is well adapted for charting the stars is a refractor, constructed on the lines of the international photo-telescopes, but it remains to be proved whether the instrument which was designed by Professor Pickering, and now constructed in America, will give better results. My reasons for preferring refractors with moderate apertures and focal length, such as are the international instruments, are the comparative

D

ease with which the adjustments can be made and maintained, the simplicity of the mechanical operations, and the perfection of the star images on the negatives.

If I were choosing an instrument for the purpose of photographing the sun, moon, and planets, I should choose a refractor, with an objective of six or eight inches aperture and very great focal length; but for the delineation of faint stars and faint nebulosity, I give preference to the silver-on-glass reflector, and should choose an instrument of twenty-seven inches clear aperture, and eleven feet three inches focal length, the mirror being figured, and every part of the mechanism made in the most perfect manner possible. The photographs are to be taken in the focus of the mirror, without being reflected from a plane, the photo-plate with its carrier being placed inside the tube of the telescope.

This plan has the merit of utilizing the maximum photo-action, by the light of celestial objects upon the sensitive film, over a field of about two degrees in diameter. A lesser field than four square degrees would not contain some of the known nebulæ, and a larger field would not show the star discs near its boundary sufficiently free from distortion.

The focal length of this ideal reflector would be equal to that of the international photo-telescopes, and consequently the photo-plates used would be interchangeable, and be easily comparable with those taken by the international instruments.

A guiding telescope, having an objective of ten inches in diameter, should be attached to the tube of the photo-telescope, for the purpose of applying corrections to small clock errors and refraction. It should be provided with a rectangular, double-sliding eye-piece, with cross-wires made to traverse the whole of the field of the objective, and the slow motion gearing, both in Right Ascension and Declination, should be so finely made that a motion of two seconds of arc would be palpable to the touch. The guiding telescope, when it is applied for making corrections during long exposures of the plates, is liable, together with the photo-telescope, to introduce displacements of the stellar images, by reason of strains, flexure. and slipping of the axes The rolling of the mirror, or of the objective, or of both independently in their respective cells, when the instrument is moved over a large arc from one side of the meridian to the other, are also serious causes of displacements of the stellar images.

I have recently discovered a method by which these troublesome occurrences can be effectively eliminated from the reflector. The method is as follows :—

COLLIMATION OF THE MIRROR.

My photo-mirror is pierced at its centre with a hole of about three inches in diameter, and a similar hole coincident with it is formed in the bottom of the mirror cell. A small telescope, having a magnifying power of about seventy diameters, with cross-wires fixed in its eyepiece, is firmly fixed with adjusting screws to the back of the cell, and pointed through the hole in the cell and mirror towards the focus of the mirror, where the photo-plate is fixed. On the back of the shutter of the slide, which carries the photo-plate, is fixed a silvered glass plane, and when the slide has been placed in its supports for taking a photograph, its position is of course parallel with the photo-mirror, and when a star image is formed upon the silvered plane by the mirror, the cross-wires in the eyepiece of the small telescope are directed so as to bisect the star image. Then the cross-wires in the eyepiece of the guiding telescope attached to the reflector are adjusted so as to bisect the same star, and thus, the star image being bisected by the cross-wires in the oculars of both telescopes, the adjustments are complete.

The silvered plane is withdrawn by the act of opening the shutter, which covers the photo-plate until the exposure is to be commenced.

The corrections, which have to be applied during an exposure of a plate, are made by keeping the star image bisected by the cross-wires in the ocular of the guiding telescope, and occasionally, during a long exposure, the small telescope and silvered plane are brought into action to test the coincidence or otherwise of the collimation of the guiding telescope with that of the photo-mirror. If a discrepancy is found to exist between them, the coincidence is restored by bringing, with the Right Ascension and Declination slow motion gearing, the star image first to bisection by the cross-wires of the small telescope attached to the mirror cell, and then readjusting the cross-wires of the guiding telescope upon the star in its altered position.

This double check ought not to be required oftener than once an hour during an exposure in a well-made photo-telescope, but the method gives command of the reflector in photographic work, which makes it at least equal in accuracy of performance to the refractor. The method also is useful in collimating the photo-mirror, and the small telescope will be designated the *collimating telescope*.

There is another method adopted by Dr. Common* for correcting the errors when large reflectors are used in photography, which consists in the arrangement of the slides, which hold the photographic plates, upon a frame having two motions at right angles

---

* "Monthly Notices, R.A.S.," Vol. 49, p. 297, 1888-9.

with each other. A microscope is attached to the carrier slide of the photo-plate, and by constantly watching a star image, when placed on the cross-wires of a microscope, and moving the screws of the sliding frame when a correction is required, circular star discs can, with care, be obtained, though the labour is fatiguing.

Whether the refractor or the reflector be the instrument chosen, the following conditions are essential in either of them, when long exposures of the plates are contemplated; and as long exposures will undoubtedly be necessary to enable us materially to advance the knowledge of astronomical phenomena, I shall not, without sufficient reason, assert that satisfactory scientific work cannot be executed if these conditions are disregarded.

ESSENTIALS OF A PHOTO-TELESCOPE.

1st. The stand of the telescope must be strong and firm, and fixed so that it will not vibrate when a moderately high wind is blowing on the telescope, nor show vibration when the operator is jumping on the floor of the observatory, or when lightly tapping the stand with the hand.

2nd. The tube of the telescope must be rigid, and not liable to flexure to an extent greater than two seconds of arc, when it is weighted with the cell, the mirror, or the objective, and all accessories attached, even though the tube should be moved in azimuth, or in declination, over an arc of seventy-five degrees in amplitude. This limit of error must also include any slipping, or looseness, in the polar and declination axes; any flexure in the cell, and rolling of the mirror or object glass in the cell; any slipping of the plate-holder and plates, or of the attachments of the photo-plate holder to the tube. In short, I consider that two seconds of arc should include all deviations which would cause perceptible distortion of the star images on a photograph.

3rd. The driving clock, sector and tangent screw, must be so perfectly made, that when a sixth magnitude star near the meridian is viewed on the cross-wires in the eyepiece of the guiding telescope, it shall be kept steadily bisected by the cross-wires for consecutive intervals of at least three minutes each, under a magnifying power of one hundred and fifty diameters.

The same care and mechanical skill should be exercised in the construction of the complete photo-telescope as in the construction of a good sidereal clock, the time to be kept by both instruments requiring similar accuracy. But the mechanician has to provide in one case a clock movement involving a weight of a few ounces to be moved with accuracy, whereas, in the other case, the weight to be moved may be a ton or more, and equal accuracy in the result is required in both instruments.

It may be stated as an axiom that any mechanism that will keep a star steadily bisected on the cross-wire in the ocular of the guiding telescope during consecutive intervals of three minutes of time, either with or without a pendulum and electrical control, provides all that is essential for doing good photographic work.

Very frequent application of the eye to the guiding telescope will, even with a perfectly constructed mechanism, be necessary, because of atmospheric disturbances, the chief of which is the frequent changes in refraction during an exposure. These occur unexpectedly, at any moment, by reason of currents of air of different density and temperature passing over the field which is being photographed.

METHOD FOR TESTING THE STABILITY OF A PHOTO-INSTRUMENT.

Place in the photo-telescope a sensitized plate, in readiness for making an exposure; then select a star of about third magnitude, and distant about two hours *east* of the meridian, and at any altitude between the zenith and the equator. Bring the cross-wires of the guiding telescope, by the Right Ascension and Declination slow-motion handles, to bisect the star image. A photograph of the star, with an exposure of about two minutes, should then be taken, and the slide closed.

Move the photo-telescope, so as to place the cross-wires again upon another star, which is on or near the meridian, and at any altitude; then expose the photographic plate for two minutes as before, and close the slide.

Move the telescope, and place it upon a star which is about two hours *west* of the meridian, and again bring the cross-wires in the ocular of the guiding telescope to bisect the star image. Expose the plate for the third time, during two minutes, upon the star.

Now, if after developing the plate it is found that the three stars which have been exposed are superimposed, so that only one circular stellar image is formed on the photograph, it may be concluded that all the parts of the photo-instrument are in proper order for photographic work, requiring exposures of the plates up to four hours duration; but if the composite stellar image is not round, but distorted, it would be useless attempting exposures of four hours till the mechanism is brought to the necessary degree of stability. The weak places must therefore be found out and remedied, and the testing be repeated time after time, until the adjustments of the instrument as a whole can be reliably maintained during long exposures of the sensitized plates.

PHOTOGRAPHIC PLATES: THEIR EXPOSURES AND DEVELOPMENT.

The plates most suitable for photographing the Sun, Moon and the planets, are the slow emulsion bromide plates, upon which the development of the images can be watched so as to give the required density and details; but for photographing nebulæ and faint stars, the most rapid plates that can be prepared are required, and with these, exposures of many hours duration will have to be given, in order to show very faint stars and faint nebulosity.

The photo-images of the greater number of the nebulæ and stars are so faint and small that they cannot be seen during the development of the plates; we have, therefore, to rely upon the colour of the film to determine the required development. As a rule, the development should be carried on until the film is of a dark colour when viewed in the developing solution in a dish, but it should not be allowed to darken the film to a degree that prints cannot be taken from the negative.

The "extra rapid" bromide plates which are prepared by various makers, and designated by a high sensitometer number, are found, when tested, to differ largely in sensitiveness, though marked with the same number, and this is frequently the case in plates which have been prepared with care by the same maker. The photo-astronomer should, therefore, carefully test each parcel of plates he receives before using them in his work.

Another source of annoyance arises in consequence of due care not being exercised by the makers of photo-plates, in the selection of glass free from air bubbles, scratches, and striæ. These defects apply most pointedly to plates made in foreign countries, and I have several times found photographs, which had been exposed in the telescope for some hours, spoilt in consequence of bubbles in the plates.

Plate 3.

GREAT NEBULA IN ANDROMEDA.

Great Nebula in Andromeda.

R.A. 0h. 37m. 17s. ; Dec. N. 40° 43·4′

The photograph covers the region between R.A. 0h. 32m. 50s. and R.A. 0h. 41m. 43s. Declination between 39° 40′ and 41° 40′ North.

Scale—1 millimetre to 30 seconds of arc.

Co-ordinates of the Fiducial stars marked with dots for the epoch A.D. 1900.

Star (.) D.M. No. 131—Zone+40°	...	R.A. 0h. 33m. 1·2s.	...	Dec. N. 40° 26·4′	...	Mag. 8·5
„ (··) „ „ 158 „ 39°	...	„ 0h. 36m. 35·5s.	...	„ 40° 5·5′	...	„ 7·0
„ (∵) „ „ 154 „ 40°	...	„ 0h. 39m. 21·0s.	...	„ 40° 45·7′	...	„ 9·0
„ (∷) „ „ 158 „ 40°	...	„ 0h. 40m. 33·6s.	...	„ 40° 15·6′	...	„ 7·5

The photograph was taken with the 20-inch reflector on December 29th, 1888, between sidereal time 1h. 38m. and 5h. 45m., with an exposure of the plate during four hours.

Three nebulæ are shown on the photograph. 1st, the Great Nebula, Messier 31; 2nd, the one on the *south* side is h 51 (M. 32), and was discovered by Le Gentil in 1749; 3rd, that on the *north preceding* side is h 44, and was discovered by Caroline Herschel in 1783.

REFERENCES.

A monograph of this nebula was prepared by George P. Bond, and published in the *Memoirs of the American Academy of Arts and Sciences, New Series*, Vol. III., pp. 75 to 86, 1848. The monograph is accompanied by a carefully prepared and well executed drawing of the nebula, and, taken together, they contained the fullest information extant concerning the object, until the photographs, of which the annexed is one of a series, were taken.

Bond, with the 15-inch refractor of the Harvard Observatory, saw in the nebula two dark bands or canals, perfectly straight, suddenly terminated, and slightly diverging at an angle of about three degrees. He also states that he estimated above 1500 stars to be involved within the limits of the nebula.

A photograph which I took with the 20-inch reflector on October 10th, 1887, revealed for the first time the true character of the Great Nebula, and one of the features exhibited was that the dark bands, referred to by Bond, formed parts of divisions between symmetrical rings of nebulous matter surrounding the large diffuse centre of the nebula. Other photographs were taken in 1887, November 15th ; 1888, October 1st ; 1888, October 2nd ; 1888, December 29th ; besides several others taken since, upon all of which the rings of nebulosity are identically shown, and thus the photographs confirm the accuracy of each other, and the objective reality of the details shown of the structure of the nebula.

The Photograph annexed is an enlargement from the negative taken in 1888, December 29th, and will remain a permanent record, unquestionable in accuracy, of the state or the appearance of the three nebulæ shown on the night when the photograph was taken, so that any changes that may take place in the form or density of the nebulosity, or in the positions or magnitude of the stars, will in future be capable of demonstration.

These photographs throw a strong light on the probable truth of the *Nebular Hypothesis,* for they show what appears to be the progressive evolution of a gigantic stellar system. Much additional evidence of a similar confirmatory character will also be seen on examination of other photographs of nebulæ which are given in the pages following.

Plate 4.

CHART OF THE REGION OF GAMMA CASSIOPEIÆ.

CHART OF THE REGION OF GAMMA CASSIOPEIÆ.

Chart of the Region of Gamma Cassiopeiæ.

R.A. 0h. 50m. 40s. ; Dec. N. 60° 10·6′.

The chart covers the region between R.A. 0h. 45m. 0s. and R.A. 0h. 55m. 57s. Declination between 59° 21′ and 61° 2′ North.

Scale—1 millimetre to 24 seconds of arc.

Co-ordinates of the Fiducial stars marked with dots for the epoch A.D. 1900.

Star (.) D.M. No. 124—Zone+60°	...	R.A. 0h. 47m. 7·5s.	...	Dec. N. 60° 33·9′	...	Mag. 5·0
„ (··) „ „ 146 „ 59°	...	„ 0h. 50m. 45·6s.	...	„ 59° 49·3′	...	„ 6·3
„ (∵) „ „ 137 „ 60°	...	„ 0h. 51m. 16·1s.	...	„ 60° 53·1′	...	„ 7·0
„ (∷) „ „ 161 „ 59°	...	„ 0h. 53m. 57·5s.	...	„ 59° 58·3′	...	„ 7·2

The photograph was taken with the 20-inch reflector on January 17th, 1890, between sidereal time 3h. 13m. and 4h. 45m., with an exposure of the plate during ninety minutes.

The star *Gamma* is involved, and is distinctly visible on the negative, in the centre of the nebulous patch of light shown on the photograph, which is probably caused by the illumination of a part of the earth's atmosphere by the star, for there is no conclusive evidence to show that the star itself is involved in nebulosity.

Admiral Smyth* states that *Gamma* is a bright star with a distant telescopic companion of 13th magnitude, the position angle being 347°, and distance 350″, for the epoch 1837·68. The photographic negative shows that there are 57 stars, between 13th and 16th magnitude, within a circle of 350 seconds of arc radius of *Gamma*, and that the nearest of them is 83 seconds of arc off; but it does not appear that any one of them is a companion, though several of them are of 13th to 14th magnitude, and much nearer to the primary than the companion indicated by Admiral Smyth.

---

* "Cycle of Celestial Objects." Chambers' edition, 1881, p. 22.

E

Plate 5.

CLUSTER MESSIER 103 CASSIOPEIÆ.

Cluster Messier 103 Cassiopeiæ.

R.A. 1h. 26m. 37s.; Dec. N. 60° 10·9'.

The photograph covers the region between R.A. 1h. 21m. 30s. and R.A. 1h. 32m. 24s. Declination between 59° 19' and 61° 0' North.

Scale—1 millimetre to 24 seconds of arc.

Co-ordinates of the Fiducial stars marked with dots for the epoch A.D. 1900.

Star (.)	D.M. No. 260—Zone + 59°	...	R.A. 1h. 23m. 18·5s.	...	Dec. N. 59° 44·3'	...	Mag. 7·2
„ (··)	„ „ 255 „ 60°	...	„ 1h. 25m. 3·9s.	...	„ 60° 31·7'	...	„ 7·7
„ (∵)	„ „ 269 „ 59°	..	„ 1h. 26m. 32s.	...	„ 59° 56·1'	...	„ 7·5
„ (∷)	„ „ 281 „ 60°	...	„ 1h. 29m. 54·3s.	...	„ 60° 45·6'	...	„ 8·5

The photograph was taken with the 20-inch reflector on December 23rd, 1892, between sidereal time 2h. 43m. and 3h. 43m., with an exposure of the plate during sixty minutes.

REFERENCES.

N.G.C. 581.   G.C. 341.   h 126.

Sir J. Herschel describes the cluster as bright, round, rich, pretty large; stars 10th to 11th magnitude.

The Photograph shows the stars down to about the 15th magnitude, and several of them have what appear to be *comites* at a distance of a few seconds of arc. It is a remarkable feature, to be seen on almost all photographs which have been exposed for sixty minutes and upwards, that the bright stars have one or more small stars like *comites* in close proximity to them, and the correlation of photographs taken at intervals of a few years will readily, and with certainty, demonstrate the connection or otherwise of the faint stars with their supposed primaries.

The cluster is not strikingly round, or symmetrical, nor are the stars very close together, and (unlike globular clusters) there is no appearance of nebulosity in it.

There is a remarkable group of eight stars, all of about 10th magnitude, arranged in two parallel straight lines with their photo-discs partly overlapping, so that they appear

E 2

like a star trail.   They are about forty minutes of arc distant from the cluster, in the *north following* quadrant, and are surrounded by stars of 10th to 14th magnitude.

Groups of multiple stars are also of frequent occurrence on many of the photographs, and their photo-discs generally overlap, thus forming an irregular figure. On the photograph annexed are several instances, and the deviations from the circular form of the star images are (excepting where distortion occurs) due to close double or multiple stars.

Plate 6.

NEBULÆ MESSIER 76 & ♓ I 193 PERSEI.

Nebulæ M. 76 and ♄. I. 193 Persei.

R.A. 1h. 36m. 0s.; Dec. N. 51° 3·9′.

The photograph covers the region between R.A. 1h. 33m. 20s. and R.A. 1h. 38m. 44s. Declination between 50° 32′ and 51° 35′ North.

Scale—1 millimetre to 15 seconds of arc.

Co-ordinates of the Fiducial stars marked with dots for the epoch A.D. 1900.

Star	D.M. No.	Zone	R.A.	Dec.	Mag.
( )	363	+51°	1h. 34m. 1·6s.	N. 51° 21·4′	8·6
(··)	326	50°	1h. 35m. 7·9s.	51° 2·5′	9·0
(∵)	336	50°	1h. 37m. 18·7s.	51° 1·1′	7·0
(∷)	339	50°	1h. 38m. 6·9s.	50° 47·3′	9·5

The photograph was taken with the 20-inch reflector on November 7th, 1891, between sidereal time 2h. 2m. and 3h. 2m., with an exposure of the plate during one hour.

REFERENCES.

N.G.C. 650 and 651.   G.C. 385 and 386.   ♄. I. 193.

Sir J. Herschel, in the G.C., describes both the nebulæ as very bright; one preceding the other and forming a double nebula.

Lord Rosse (*Obs. of Neb. and Cl., p.* 21) describes it as a new spiral, consisting principally of two bright knots. The nebulosity terminates very suddenly on the south edge, where there is a star. The descriptive matter is illustrated by a sketch.

The Photograph shows the two nebulæ to be one only, with patches of dense nebulosity near the *south preceding* and *north following* margins, and there is a sharply defined star of about 14th magnitude in the centre of figure and also a star of about 12th magnitude connected with the nebula at the *south following* margin. There are faint, star-like patches of nebulosity distributed over the nebula, and two very faint, widely detached patches on the *south following* and *north preceding* sides. The figure of the nebula suggests that it is a broad ring seen edgewise.

Plate 7.

CLUSTERS H VI 33, 34 PERSEI.

Clusters Ḧ. VI. 33, 34 Persei.

R.A. 2h. 12m. 2s. and 2h. 15m. 23s.; Dec. N. 56° 41·3′ and 56° 39·2′.

The photograph covers the region between R.A. 2h. 8m. 25s. and R.A. 2h. 18m. 20s. Declination between 55° 50′ and 57° 32′ North.

Scale—1 millimetre to 24 seconds of arc.

Co-ordinates of the Fiducial stars marked with dots for the epoch A.D. 1900.

Star	D.M. No.	Zone	R.A.	Dec. N.	Mag.
(·)	471	+56°	2h. 9m. 52·2s.	56° 35·4′	6·6
(··)	486	56°	2h. 11m. 1·9s.	57° 3·4′	6·5
(∵)	567	56°	2h. 14m. 49·9s.	56° 26·8′	8·4
(∷)	593	56°	2h. 15m. 54·7s.	56° 55·9′	7·0

The photograph was taken with the 20-inch reflector on January 13th, 1890, between sidereal time 2h. 27m. and 5h. 47m., with an exposure of the plate during three hours.

REFERENCES.

N.G.C. 869 and 884. G.C. 512 and 521. _h_ 207 and 212. Ḧ VI. 33 and 34.

Sir J. Herschel, in the G.C., describes the clusters as very large; very rich; stars 7th to 14th; ruby star in the middle of Cl. 34.

Lord Rosse (_Obs. of Neb. and Cl., pp._ 26, 27) describes some red, yellow, and blue stars in the clusters, and, generally, the appearances when viewed in the three-foot and six-foot telescopes.

The Photograph presents to the eye the stars in the two clusters and in the surrounding parts of the sky with a completeness and accuracy of detail never before seen. The stars are shown in their true relative positions and magnitude to about the 16th, and among them are many apparent double, triple, and multiple stars. They also appear to be arranged in clusters, curves, festoons and patterns that are suggestive of some physical connection existing between the groups, but it is premature to assert that these appearances are not due to perspective effect by the eye arranging numerous close points of light into various patterns. Similar photographs to this, taken at intervals of several years between them, will determine the reality, or otherwise, of these remarkable groupings of the stars.

These clusters differ from others, such as the *Pleiades, M.* 13 *Herculis and M.* 15 *Pegasi,* inasmuch as they are free from any trace of nebulosity, and this fact suggests the probability of the absorption of the nebulosity by the stars in these two clusters, and that similar absorption of the nebulosity may be in progress in those cluterss in which the photographs show its existence.

Plate 8.

NEBULA I♭ V 19 ANDROMEDÆ.

Nebula ♄. V. 19 Andromedæ.

R.A. 2h. 16m. 15s.; Dec. N. 41° 53·6'.

The photograph covers the region between R.A. 2h. 12m. 41s. and R.A. 2h. 19m. 55s. Declination between 41° 7' and 42° 48' North.

Scale—1 millimetre to 24 seconds of arc.

Co-ordinates of the Fiducial stars marked with dots for the epoch A.D. 1900.

Star (.) D.M. No. 440—Zone +41°	...	R.A. 2h. 13m. 12·0s.	...	Dec. N. 41° 37·9'	...	Mag 9·1
,, (··) ., ,, 502 ,, 42°	...	., 2h. 14m. 46·5s.	...	,, 42° 29·7'	..	,, 6·5
,, (·.·) ,, ,, 509 ,, 42°	...	,, 2h. 16m. 53·1s.	...	,, 42° 33·3'	...	,, 8·1
., (::) ,, ,, 453 ,, 41°	...	,, 2h. 17m. 45·8s.	...	,, 41° 38·9'	...	,. 6·9

The photograph was taken with the 20-inch reflector on December 21st, 1891, between sidereal time 2h. 26m. and 5h. 33m., with an exposure of the plate during three hours.

REFERENCES.

N.G.C. 891.   G.C. 527.   h 218.   ♄ V. 19.

Sir J. Herschel, in the G.C., describes the nebula as bright; very large; very much extended 22° 3'; and in his *Obs. of Neb. and Cl.*, p. 498, he states that it is of the last degree of faintness, and there can hardly be a doubt of its being a thin, flat ring of enormous dimensions, seen very obliquely.   A drawing of the nebula is given in the *Phil. Trans.*, 1833, *pl. X., fig.* 28, which agrees well with the photograph.

Lord Rosse, in the *Phil. Trans.*, 1861, *p.* 712, describes the nebula and gives a sketch of it, showing its general outlines and the positions of five stars which are involved in it. The stars indicated are in fair agreement with those shown on the photograph.

In the *Obs. of Neb. and Cl.*, *p.* 28, Lord Rosse records five observations of the nebula between 1850 and 1874, but they do not accord so well with the photograph as those referred to in the foregoing paragraph.

The descriptions given by the observers just cited will also apply to the photograph, and the suggestion by Sir J. Herschel that the nebula is a thin, flat ring of enormous dimensions, seen very obliquely, receives strong confirmation.

Plate 9.

CLUSTER MESSIER 34 PERSEI.

Cluster Messier 34 Persei.

R.A. 2h. 35m. 35s.; Dec. N. 42° 21·0'.

The photograph covers the region between R.A. 2h. 31m. 45s. and R.A. 2h. 39m. 8s. Declination between 41° 30' and 43° 12' North.

Scale—1 millimetre to 24 seconds of arc.

Co-ordinates of the Fiducial stars marked with dots for the epoch A.D. 1900.

Star (.) D.M. No. 566—Zone +42°	...	R.A. 2h. 33m. 23·4s.	...	Dec. N. 42° 33·6'	...	Mag. 8·8
„ (··) „ „ 514 „ 41° ..		„ 2h. 35m. 5·9s.	...	„ 42° 9·9'	...	„ 8·0
„ (··) „ „ 615 „ 42° ...		„ 2h. 37m. 6·3s.	...	„ 42° 12·1'	...	„ 8·5
„ (::) „ „ 614 „ 42° ..		„ 2h. 37m. 5·9s.	...	., 43° 6·7'	...	„ 6·8

The photograph was taken with the 20-inch reflector on December 26th, 1892, between sidereal time 1h. 29m. and 2h. 29m., with an exposure of the plate during one hour.

REFERENCES.

N.G.C. 1039.   G.C. 584.   h 248.

Sir J. Herschel, in the G.C., describes the cluster as bright; very large; little compressed; scattered stars, 9th magnitude.

Lord Rosse (*Obs. of Neb. and Cl., p.* 31) describes it as a scattered cluster; an isosceles triangle of stars on the *north* side, points *south* to a red star which has a small companion *south preceding.*

The Photograph shows the stars down to about the fifteenth magnitude, and the descriptions given by the observers cited above will, in the absence of drawings, explain parts of the cluster.

Plate 10.

NEBULA MESSIER 77 CETI.

Nebula Messier 77 Ceti.

R.A. 2h. 37m. 34s.; Dec. S. 0° 26·3′.

The photograph covers a region 19 minutes of arc in diameter, with the nebula in the centre of the circle.

Scale—1 millimetre to 6 seconds of arc.

Co-ordinates of the nebula for the epoch A.D. 1900. R.A. 2h. 37m. 34·6s.; Dec. S. 0° 26·7′

The photograph was taken with the 20-inch reflector on November 26th, 1892, between sidereal time 2h. 34m. and 4h. 4m., with an exposure of the plate during ninety minutes.

REFERENCES.

N.G.C. 1068. G.C. 600. *h* 262.

Sir J. Herschel, in the G.C., describes the nebula as very bright; pretty large; irregularly round; suddenly brighter in the middle; partly resolved; with a nucleus. Star 130° 2′.

Lord Rosse (*Obs. of Neb. and Cl., p.* 32) records twelve observations of the nebula made between the years 1848 and 1874 and considered it to be a spiral, with a sharp nucleus very little elongated *north* and *south.* A drawing is given in the *Phil. Trans. for* 1861, *pl. XXV., fig.* 6, and another in the *Obs. of Neb. and Cl., pl.* 1.

Lassell, in the *Memoirs of the Royal Astronomical Society, Vol. XXXVI., pl.* 1., *fig.* 2, *p.* 40, describes the nebula as a spiral with three stars involved, and a nebulous star near it.

The Photograph shows the nebula to have a stellar nucleus with projecting *ansæ* of dense nebulosity on the *north following* and *south preceding* sides, and surrounding the *ansæ* is a zone of faint nebulosity surrounded by a broad nebulous ring, which is studded with strong condensations resembling stars with irregular margins. Six or eight of the nebulous condensations can, on the negative, be discerned, and the nebula is strongly suggestive of the idea that a stellar or planetary system is actually now in process of formation.

The nebulous star, to which Lassell refers, does not, on the photograph, show any nebulosity around it.

Plate 11.

NEBULÆ IN THE PLEIADES.

NEBULÆ IN THE PLEIADES.

Nebulæ in the Pleiades.

The photograph covers the region between R.A. 3h. 38m. 7s. and R.A. 3h. 44m. 3s. Declination between 23° 1′ and 24° 43′ North.

Scale—1 millimetre to 24 seconds of arc.

Co-ordinates of the Fiducial stars marked with dots for the epoch A.D. 1900.

Star (.) D.M. No. 505—Zone +23°	...	R.A. 3h. 38m. 52·1s.	...	Dec. N. 23° 58·8′	...	Mag. 6·5
„ (··) „ „ 562 „ 24°	...	„ 3h. 41m. 1·8s.	...	„ 24° 12·8′	...	„ 7·5
„ (∵) „ „ 537 „ 23°	...	„ 3h. 41m. 24·9s.	...	„ 23° 29·5′	...	„ 7·5
„ (∷) „ „ 561 „ 23°	...	„ 3h. 43m. 24·7s.	...	„ 24° 4·6′	...	„ 7·5

The photograph was taken with the 20-inch reflector on December 8th, 1888, between sidereal time 2h. 44m. and 6h. 52m., with an exposure of the plate during four hours.

Three photographs of the *Pleiades* were taken with the 20-inch reflector between October 23rd, 1886, and the end of December following, upon each of which nebulosity in the group, of the same character and extent, was shown as that visible on the photograph annexed, and my references to them are published in the *Monthly Notices of the R.A.S., Vol. XLVII., pp.* 24 *and* 90; *Vol. XLIX., p.* 120.

Mr. W. H. Wesley (*Asst. Sec. R.A.S.*) has written a concise history (accompanied by four illustrative drawings) of the discovery of nebulæ in the *Pleiades*, which was published in the journal of the *Liverpool Astronomical Society, Vol. V., pp.* 148-150.

The photo-images of the bright stars are obscured on the prints by the density of the nebulosity around them, to which is added the photo-action of the light of the stars upon our atmosphere, but on the negatives the stellar images are very clearly seen in the midst of the surrounding nebulosity.

The projecting spikes of light that are shown round the bright stars are due to instrumental causes. They are the effects of interference by the supports of the photographic plate-holder (within the tube of the telescope) with the light in its passage to the speculum.

Plate 12.

REGION OF NOVA AURIGÆ.

Region of Nova Aurigæ.

The photograph covers the region between R.A. 5h. 20m. 4s. and R.A. 5h. 26m. 25s. Declination between 29° 29′ and 31° 10′ North.

Scale —1 millimetre to 24 seconds of arc.

Co-ordinates of the Fiducial stars marked with dots for the epoch A.D. 1900.

Star (.)	D.M.	No. 898—Zone+30°	...	R.A.	5h. 20m. 44·0s.	...	Dec. N.	30° 6·9′	...	Mag.	6·2
„ (··)	„	„ 912 „ 30°	...	„	5h. 23m. 20·4s.	...	„	30° 30·9′	...	„	8·5
„ (∴)	„	„ 911 „ 29°	...	„	5h. 23m. 44·2s.	...	„	29° 28·1′	...	„	7·5
„ (∷)	„	„ 923 „ 30°	...	„	5h. 25m. 26·9s.	...	·,	30° 50·8′	...	„	8·9
Nova (N)	„	„ ...	„	...	...	„	5h. 25m. 34s.	...	„	30° 21′	...	„	...

The photograph was taken with the 20-inch reflector on February 18th, 1892, between sidereal time 5h. 2m. and 8h. 8m., with an exposure of the plate during three hours. Twelve other photographs were taken between February 5th, 1892, and January 20th, 1893, with exposures of the plates during intervals varying from five minutes to 110 minutes.

My references to these photographs are published in the *Monthly Notices of the R.A.S., Vol. LII., pp.* 371, 372, *and Vol. LIII., pp.* 123, 124.

Nova Aurigæ was discovered by Dr. T. D. Anderson at Edinburgh, and announced by him on February 1st, 1892. It excited great interest, and was carefully observed by astronomers until its light diminished so as to be too feeble for general observations. The records concerning it are numerous, and will be found in the *Monthly Notices, Vol. LIII., pp.* 269, 270, and other astronomical publications, issued between February, 1892, and the beginning of 1893.

Plate 13.

CLUSTER MESSIER 36 AURIGÆ.

Cluster Messier 36 Aurigæ.

R.A. 5h. 29m. 42s.; Dec. N. 34° 4·3′.

The photograph covers the region between R.A. 5h. 26m. 31s. and R.A. 5h. 33m. 6s. Declination between 33° 16′ and 34° 57′ North.

Scale—1 millimetre to 24 seconds of arc.

Co-ordinates of the Fiducial stars marked with dots for the epoch A.D. 1900.

Star (.) D.M. No. 1088—Zone + 34°	...	R.A. 5h. 27m. 34·1s.	...	Dec. N. 34° 33·7′	...	Mag. 8·5
,, (··) ,, ,, 1087 ,, 33°	...	,, 5h. 28m. 37·2s.	...	,, 33° 53·5′	...	,, 8·5
,, (·:) ,, ,, 1118 ,, 34°	...	,, 5h. 30m. 31·3s.	...	,, 34° 45·0′	...	,, 8·1
,, (::) ,, ,, 1103 ,, 33°	...	,, 5h. 31m. 13·7s.	...	,, 33° 54·1′	...	,, 8·3

The photograph was taken with the 20-inch reflector on February 8th, 1893, between sidereal time 3h. 55m. and 5h. 25m., with an exposure of the plate during ninety minutes.

REFERENCES.

N.G.C. 1960.   G.C. 1166.   h 358.

Sir J. Herschel, in the G.C., describes it as bright; very large; very rich; little compressed; stars 9th to 11th scattered.

Lord Rosse (*Obs. of Neb. and Cl., p.* 48) describes it as a coarse cluster; stars large; fine double star in it.

The Photograph is in general agreement with the descriptions cited, but it will be observed that several of the stars are apparently double and triple, and there is at least one multiple star.   There is no nebulosity in the cluster, and the stars are shown to about the 16th magnitude.

Plate 14.

"CRAB" NEBULA IN TAURUS.

"CRAB" NEBULA IN TAURUS.

"Crab" Nebula in Taurus.

R.A. 5h. 28m. 30s.; Dec. N. 21° 57′.

The photograph covers the region between R.A. 5h. 25m. 34s. and R.A. 5h. 31m. 26s. Declination between 21° 7′ and 22° 48′ North.

Scale—1 millimetre to 24 seconds of arc.

Co-ordinates of the Fiducial stars marked with dots for the epoch A.D. 1900.

Star (.)	D.M. No.	939—	Zone+22°	...	R.A.	5h. 27m. 7·3s.	...	Dec. N.	22° 13·3′	..	Mag.	9·5
,, (··)	,,	,,	892	,,	21°	...	,,	5h. 28m. 36·8s.	...	,,	21° 33·1′	...	,,	9·0
,, (·.)	,,	,,	952	,,	22°	...	,,	5h. 29m. 23·3s.	...	,,	22° 22·5′	...	,,	8·5
,, (::)	,,	,,	902	,,	21°	...	,,	5h. 30m. 26·9s.	...	,,	21° 55·7′	...	,,	7·4

The photograph was taken with the 20-inch reflector on February 2nd, 1892, between sidereal time 6h. 0m. and 9h. 7m., with an exposure of the plate during three hours.

REFERENCES.

N.G.C. 1952.   G.C. 1157.   h 357.

Sir J. Herschel, in the G.C., describes the nebula as very bright; very large; extended in the direction 135°; very gradually a little brighter in the middle; mottled as with stars. He gave a sketch of it in the *Phil. Trans. for* 1833, *Pl. XVI., fig.* 81, but it does not much resemble the photograph.

Lord Rosse, in the *Phil. Trans.*, 1844, *Pl. XVIII., fig.* 81, *p.* 322, states :—" It is no longer an oval resolvable nebula; we see resolvable filaments singularly disposed, springing principally from its southern extremity and not, as is usual in clusters, irregularly in all directions."   The drawing is, in accordance with this description, an ovate body with filamentous projections around it.   There is no resemblance between these descriptions and the photograph, but in the *Obs. of Neb. and Cl., p.* 47, and the drawing and marginal sketch given, there are several features shown that approximately correspond with the photograph.

Lassell, in the *Mem. R.A.S., Vol. XXIII., p.* 59, *pl.* 2, *fig.* 1, and *Vol. XXXVI., p.* 41, *pl.* 2, *fig.* 6, describes and shows in his drawings the nebula with projecting appendages, but they are not like those shown by Lord Rosse, and there is nothing resembling them on the photograph.

The Photograph of the nebula as it appears on the negative is not symmetrical in form, and has a faint undefined, boldly indented margin, with a large projecting limb on the *south preceding* side. In general outline it is elliptical; with the major axis in *south following* to *north preceding* direction, and on the *north following* side is a large deep embayment with little nebulosity in it, and there is also a smaller bay with nebulosity partly filling it. The body of the nebula consists of dense masses of clouds with fainter areas between them, but they are too dense to print so as to be visible in detail on the enlargements, though they are clearly visible on the negative. There are sixteen stars involved in the nebulosity, which measures 340 seconds of arc in length and 260 seconds in breadth.

Plate 15.

GREAT NEBULA IN ORION.

GREAT NEBULA IN ORION.

Great Nebula in Orion.

No. 1.

The photograph covers the region between R.A. 5h. 27m. 20s. and R.A. 5h. 32m. 46s. Declination between 4° 34′ and 6° 15′ South.

Scale—1 millimetre to 24 seconds of arc.

Co-ordinates of the Fiducial stars marked with dots for the epoch A.D. 1900.

Star (.) D.M. No. 1289—Zone—5°	... R.A. 5h. 28m. 12·6s.	... Dec. S. 5° 24·6′	... Mag. 8·5
,, (··) ,, ,, 1167 ,, 4°	... ,, 5h. 29m. 27·3s.	... ,, 4° 52·9′	... ,, 7·5
,, (·:) ,, ,, 1334 ,, 5°	... ,, 5h. 31m. 20·7s.	... ,, 5° 42·5′	... ,, 7·8
,, (::) ,, ,, 1342 ,, 5°	... ,, 5h. 32m. 40·9s.	... ,, 5° 0·0′	... ,, 7·8

The photograph was taken with the 20-inch reflector on December 18th, 1886, between sidereal time 7h. 8m. and 7h. 23m., with an exposure of the plate during fifteen minutes.

REFERENCES.

George P. Bond, in 1861, published a valuable monograph, accompanied by a well-executed drawing and map of this nebula; and an elaborate monograph was also prepared by Prof. Edward S. Holden and published in the *Washington Astronomical Observations for* 1878, *Appendix I.* The latter monograph is comprised in 230 pages of printed matter, and is largely illustrated with copies of drawings of the nebula which were made by the various observers. The frontispiece is a copy of Bond's drawing, and near the end is a reproduction of a photograph taken by Prof. Henry Draper on March 14th, 1882.

Dr. Common, on January 30th, 1883, took a photograph of the nebula with his 3-foot reflector, and it shows the nebulosity more clearly, and extending beyond that shown on Prof. Draper's photograph.

Elaborate drawings of the nebula have been made by Lord Rosse, Lassell, Bond and others, and are described and illustrated in Holden's monograph.

The Photographs here given are three in number, and when correlated they show the structure, details and extent of the nebula with an accuracy that was never before attainable, and for the first time we have, by these photographs, including the negatives, reliable records of every detail that existed and was capable of affecting the sensitized film on the day and hour given in the text, when the photographs were taken. It will be observed on the annexed print that the stars which form the *Trapezium* are shown as one figure, though on the negative they are shown as four separate stars with their images touching. The star surrounded with nebulosity, to the *north* of the Great Nebula, is *Messier* 43.

Plate 16.

GREAT NEBULA IN ORION.

GREAT NEBULA IN ORION.

Great Nebula in Orion.

No. 2.

The photograph covers the region between R.A. 5h. 27m. 20s. and R.A. 5h. 32m. 46s. Declination between 4° 34′ and 6° 15′ South.

Scale—1 millimetre to 24 seconds of arc.

Co-ordinates of the Fiducial stars marked with dots for the epoch A.D. 1900.

Star (.) D.M. No. 1289	—Zone—5°	...	R.A. 5h. 28m. 12·6s.	...	Dec. S. 5° 24·6′	...	Mag. 8·5
,, (··) ,, ,, 1167	,, 4°	...	,, 5h. 29m. 27·3s.	...	,, 4° 52·9′	...	,, 7·5
,, (·.·) ,, ,, 1334	,, 5°	...	,, 5h. 31m. 20·7s.	...	,, 5° 42·5′	...	,, 7·8
,, (::) ,, ,, 1342	,, 5°	...	,, 5h. 32m. 40·9s.	...	,, 5° 0·0′	...	,, 7·8

The photograph was taken with the 20-inch reflector on December 24th, 1888, between sidereal time 4h. 13m. and 6h. 4m., with an exposure of the plate during eighty-one minutes.

REFERENCES.

The references which have been cited for No. 1 *ante,* apply also to the annexed Photograph, which shows large extensions of the nebulosity beyond that visible with an exposure of fifteen minutes.

The nebula Messier 43 is shown to be joined to the Great Nebula, and forms part of it. The other nebula, *h* 1180, on the *north* side is also well developed, with its characteristic streams of dense and of faint nebulosity and involved stars.

Plate 17.

GREAT NEBULA IN ORION.

GREAT NEBULA IN ORION.

Great Nebula in Orion.

No. 3.

The photograph covers the region between R.A. 5h. 27m. 20s. and R.A. 5h. 32m. 46s. Declination between 4° 34′ and 6° 15′ South.

Scale—1 millimetre to 24 seconds of arc.

Co-ordinates of the Fiducial stars marked with dots for the epoch A.D. 1900.

Star (.) D.M. No. 1289—Zone − 5°	...	R.A. 5h. 28m. 12·6s.	..	Dec. S. 5° 24·6′	...	Mag. 8·5
„ (··) „ „ 1167 „ 4°	...	„ 5h. 29m. 27·3s. ...		„ 4° 52·9′	...	„ 7·5
„ (∴) „ „ 1334 „ 5°	...	„ 5h. 31m. 20·7s. ...		„ 5° 42·5′	...	„ 7·8
„ (∷) „ „ 1342 „ 5°	...	„ 5h. 32m. 40·9s. ...		„ 5° 0·0′	...	„ 7·8

The photograph was taken with the 20-inch reflector on February 4th, 1889, between sidereal time 4h. 15m. and 7h. 40m., with an exposure of the plate during 205 minutes.

REFERENCES.

The references which have been cited for No. 1 *ante*, apply also to this Photograph, and it will be observed that the nebulosity extends far beyond the limits shown on the two photographs preceding, which are marked Nos. 1 and 2. The nebulæ, *Messier* 42, 43, and also *Herschel* 1180, are shown to be connected with each other by faint extensions of the nebulous matter; and the central parts of the *Great Nebula*, together with the central part of *h* 1180, are, on account of the long exposure of the plate, seen to have the nebulosity so strongly developed that those parts do not on the print show the details, although on the negative every detail is distinctly visible through the dense nebulosity.

Hence the reason that three prints are required to show the details that are visible on one negative with a long exposure, but by correlation of the three prints, a fair estimate can be made of the full details, and of the relative intensity of the photographic effect of the light in different parts of the nebula, and any change that may hereafter take place in the stars or in the nebulosity can with certainty be detected.

Plate 18.

CLUSTER MESSIER 37 AURIGÆ.

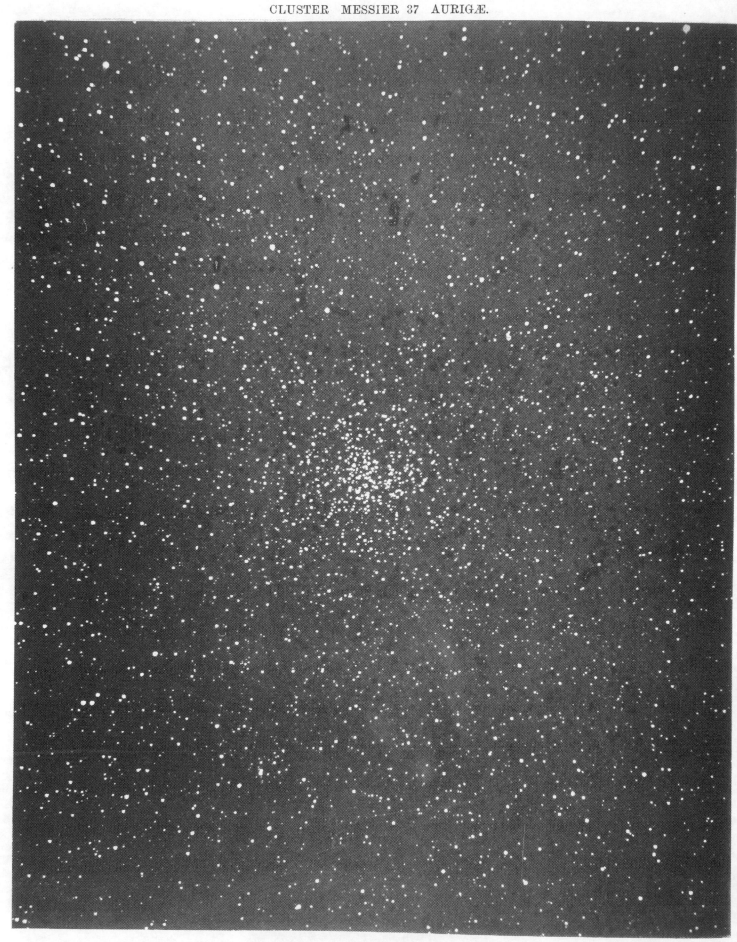

Cluster Messier 37 Aurigæ.

R.A. 5h. 45m. 45s., Dec. N. 32° 31·3′.

———

The photograph covers the region between R.A. 5h. 42m. 31s. and R.A. 5h. 48m. 58s. Declination between 31° 41′ and 33° 22′ North.

Scale—1 millimetre to 24 seconds of arc.

Co-ordinates of the Fiducial stars marked with dots for the epoch A.D. 1900.

Star (.) D.M. No. 1104—Zone + 32°	...	R.A. 5h. 43m. 51·3s.	..	Dec. N. 32° 51·0′	...	Mag. 8·8
„ (··) „ „ 1109 „ 32°	...	„ 5h. 44m. 55·3s.	...	„ 32° 6·0′	...	„ 6·6
„ (·:) „ „ 1119 „ 32°	...	„ 5h. 46m. 59·3s.	...	„ 32° 15·7′	...	„ 8·8
„ (::) „ „ 1121 „ 32°	...	., 5h. 47m. 55·8s.	...	„ 32° 29·0′	...	„ 8·7

The photograph was taken with the 20-inch reflector on February 8th, 1893, between sidereal time 5h. 39m. and 7h. 9m., with an exposure of the plate during ninety minutes.

———

REFERENCES.

N.G.C. 2099. G.C. 1295. *h* 369.

Sir J. Herschel, in the G.C., describes the cluster as rich; pretty compressed in the middle; stars large and small.

Lord Rosse (*Obs. of Neb. and Cl., p.* 51) refers to wonderful loops and curved lines of stars.

Lassell (*Mem. R.A.S., Vol. XXIII., p.* 59) describes it as a cluster of small stars pretty equal in magnitude.

The Photograph shows the cluster in a manner that can hereafter be utilized for comparison of each individual star comprised in the thousands that are recorded down to about the 16th magnitude, and it will be observed that many of them are apparently double, triple, and multiple stars.

Plate 19.

CLUSTER MESSIER 35 GEMINORUM.

Cluster Messier 35 Geminorum.

R.A. 6h. 2m. 40s. ; Dec. N. 24° 20·6′.

The photograph covers the region between R.A. 6h. 0m. 0s. and R.A. 6h. 5m. 58s. Declination between 23° 30′ and 25° 11′ North.

Scale—1 millimetre to 24 seconds of arc.

Co-ordinates of the Fiducial stars marked with dots for the epoch A.D. 1900.

Star (.) D.M. No. 1095—Zone + 24°	...	R.A. 6h. 0m. 45·1s.	...	Dec. N. 24° 19·5′	...	Mag. 8·7
„ (··) „ „ 1209 „ 23°	...	„ 6h. 2m. 4·2s.	...	„ 23° 52·5′	...	„ 8·5
„ (·.·) „ ., 1135 „ 24°	...	„ 6h. 3m. 41·2s.	...	„ 24° 13·3′	...	„ 8·4
„ (::) „ „ 1151 „ 24°	...	„ 6h. 5m. 24·5s.	...	„ 24° 26·4′	...	„ 6·7

The photograph was taken with the 20-inch reflector on February 7th, 1893, between sidereal time 6h. 4m. and 7h. 10m., with an exposure of the plate during sixty-six minutes.

REFERENCES.

N.G.C. 2168.   G.C. 1360.   *h* 377.

Sir J. Herschel, in the G.C., describes the cluster as very large ; considerably rich ; pretty compressed ; stars 9th to 16th magnitude.

Lord Rosse (*Obs. of Neb. and Cl., p.* 52) describes the cluster as magnificent, in a rich field, and he estimated 300 stars in the Finder field of 26 minutes of arc in diameter, many of which are not below 11th magnitude.

Lassell (*Mem. R.A.S., Vol. XXIII., p.* 59) describes it as a marvellously striking object.   The field of view 19′ in diameter is perfectly full of brilliant stars unusually equal in magnitude and distribution over the whole area.

The Photograph fully bears out the descriptive matter cited, and I counted 620 stars in the cluster within a circle of 26 minutes of arc in diameter, many of which have close apparent *comites*, and some are triple stars.

The cluster *H. VI.* 17., *G.C. No.* 1351, is shown on the photograph at 1m. 24s. R.A. and 0° 14′·5 Dec. *south preceding* the cluster M. 35.   It is described by Sir J. Herschel as pretty small ; much compressed ; very rich ; irregular triangle of stars extremely small.

Plate 20.

CLUSTER MESSIER 50 MONOCEROTIS.

Cluster Messier 50 Monocerotis.

R.A. 6h. 58m. 8s.; Dec. S. 8° 12·1′.

The photograph covers the region between R.A. 6h. 55m. 17s. and R.A. 7h. 0m. 46s. Declination between 7° 22′ and 9° 3′ South.

Scale—1 millimetre to 24 seconds of arc.

Co-ordinates of the Fiducial stars marked with dots for the epoch A.D. 1900.

Star (.) D.M. No. 1670—Zone—8°	...	R.A. 6h. 56m. 19·9s.	...	Dec. S. 8° 8·8′	...	Mag. 8·7
„ (··) „ „ 1695 „ 7°	...	„ 6h. 57m. 33·5s.	...	„ 7° 43·2′	...	„ 8·5
„ (∵) „ „ 1711 „ 8°	...	„ 6h. 58m. 27·2s.	...	„ 8° 21·4′	...	„ 9·0
„ (∷) „ „ 1726 „ 8°	...	„ 6h. 59m. 45·7s.	... ·	„ 8° 17·6′	...	„ 7·0

The photograph was taken with the 20-inch reflector on March 10th, 1893, between sidereal time 8h. 20m. and 9h. 50m., with an exposure of the plate during ninety minutes.

REFERENCES.

N.G.C. 2323.　G.C. 1483.　h. 425.

Sir J. Herschel, in the G.C., p. 72, describes the cluster as remarkable; very large; rich; pretty compressed; extended; stars 12th to 16th magnitude.

Lord Rosse (*Obs. of Neb. and Cl., p.* 55) records five observations between 1851 and 1876, and saw a spiral arrangement of the stars in the cluster.

The Photograph shows several of the stars in this, as in other clusters, to be apparently double and triple, but there is no strikingly spiral arrangement amongst them, though they are shown down to about the 16th magnitude.

Plate 21.

CLUSTER "PRÆSEPE" CANCRI.

CLUSTER "PRÆSEPE" CANCRI.

Cluster "Præsepe" Cancri.

The photograph covers the region between R.A. 8h. 31m. 38s. and R.A. 8h. 37m. 24s. Declination between 19° 13′ and 20° 54′ North.

Scale—1 millimetre to 24 seconds of arc.

Co-ordinates of the Fiducial stars marked with dots for the epoch A.D. 1900.

Star (.) D.M. No. 2143—Zone+20°	...	R.A. 8h. 33m. 21·7s.	...	Dec. N. 20° 1·9′	...	Mag. 8·0
„ (··) „ „ 2159 „ 20°	...	„ 8h. 34m. 26·2s.	...	„ 20° 19·4′	...	„ 7·3
„ (·.·) „ „ 2069 „ 19°	...	„ 8h. 34m. 36·5s.	...	„ 19° 42·3′	...	„ 7·0
„ (::) „ „ 2185 „ 20°	...	„ 8h. 36m. 5·3s.	...	„ 20° 13·8′	...	„ 7·5

The photograph was taken with the 20-inch reflector on February 13th, 1891, between sidereal time 8h. 1m. and 9h. 46m., with an exposure of the plate during ninety minutes.

REFERENCES,

N.G.C. 2632.   G.C. 1681.   *h* 517.   Messier 44.

The Photograph is intended to serve for a chart of the region of the *Præsepe*, and by using a lens to examine it, a large number of faint stars will be seen which, without such aid, are scarcely visible.   The bright stars in the group have much the resemblance of nebulous stars.   Their margins are notably without the sharpness of normal star images.

Plate 22.

CLUSTER MESSIER 67 CANCRI.

Cluster Messier 67 Cancri.

R.A. 8h. 45m. 46s.; Dec. N. 12° 10·6′.

The photograph covers the region between R.A. 8h. 43m. 18s. and R.A. 8h. 48m. 48s. Declination between 11° 22′ and 13° 3′ North.

Scale—1 millimetre to 24 seconds of arc.

Co-ordinates of the Fiducial stars marked with dots for the epoch A.D. 1900.

Star (.) D.M. No. 1914—Zone+12°	... R.A. 8h. 44m. 5·8s.	... Dec. N. 12° 32·3′	... Mag. 8·5
„ (··) „ „ 1923 „ 11°	... „ 8h. 44m. 50·5s.	... „ 11° 46·9′	... „ 8·9
„ (·⋅) „ „ 1925 „ 12°	... „ 8h. 45m. 57·2s.	... „ 12° 29·5′	... „ 8·0
„ (∷) „ „ 1927 „ 12°	... „ 8h. 46m. 21·9s.	... „ 12° 15·5′	... ,. 8·0

The photograph was taken with the 20-inch reflector on February 11th, 1891, between sidereal time 8h. 28m. and 10h. 4m., with an exposure of the plate during sixty-three minutes.

REFERENCES.

N.G.C. 2682.   G.C. 1712.   *h* 531.

Sir J. Herschel, in the G.C., describes the cluster as a remarkable object; very bright; very large; extremely rich; little compressed; stars 10th to 15th magnitude.

The Photograph confirms the description given, and depicts many features that could not be adequately described. In R.A. 8h. 45m., Dec. N. 12° 8′, is an assemblage (within the cluster) of five stars of about the 10th to 12th magnitude, and with their photo-images touching. If they should be proved to be physically connected the revelation would be astounding.

Plate 23.

NEBULA Iฦ I 200 LEONIS MINORIS.

Nebula H. I. 200 Leonis Minoris.

R.A. 8h. 46m. 29s. ; Dec. N. 33° 47·8′.

The photograph covers the region between R.A. 8h. 43m. 10s. and R.A. 8h. 49m. 43s. Declination between 32° 57′ and 34° 38′ North.

Scale—1 millimetre to 24 seconds of arc.

Co-ordinates of the Fiducial stars marked with dots for the epoch A.D. 1900.

Star (.) D.M. No. 1898—Zone +34°	...	R.A. 8h. 43m. 37·5s.	...	Dec. N. 34° 4·7′	...	Mag. 7·3
„ (··) „ „ 1765 „ 33°	...	„ 8h. 44m. 20·9s.	...	„ 33° 41·1′	...	„ 6·3
„ (∵) „ „ 1776 „ 33°	...	„ 8h. 47m. 40·3s.	...	„ 33° 26·1′	...	„ 8·0
„ (::) „ „ 1915 „ 34°	...	„ 8h. 48m. 53·0s.	...	„ 34° 12·5′	...	„ 9·1

The photograph was taken with the 20-inch reflector on March 19th, 1893, between sidereal time 7h. 25m. and 10h. 29m., with an exposure of the plate during three hours.

REFERENCES.

N.G.C. 2683.   G.C. 1713.   *h* 532.   H. I. 200.

Sir J. Herschel, in the G.C., p. 76, describes the nebula as very bright; very large; very much elongated 40° 9′; gradually much brighter in the middle.

Lord Rosse (*Phil. Trans.*, 1861, *p.* 718) saw the nebula slightly concave towards the *north preceding* direction; perhaps 10′ long; uncertain whether nucleus is stellar. Query, parallel dark lines exterior to nucleus as in *Andromeda*?   In the *Obs. of Neb. and Cl., p.* 70, further observations are recorded generally to the same effect.

The Photograph, in large measure, confirms the descriptions cited, and on the negative a stellar nucleus is visible in the midst of the dense central condensation.   There is no appearance of any dark lines such as are referred to, and the mottling upon the nebula is very little seen.   The extreme length of the nebula is about 8·1 minutes of arc.

Plate 24.

NEBULA ♏ I 205 URSÆ MAJORIS.

NEBULA ♏ I 205 URSÆ MAJORIS.

Nebula ♄. I. 205 Ursæ Majoris.

R.A. 9h. 15m. 6s.; Dec. N. 51° 24′.

The photograph covers the region between R.A. 9h. 10m. 43s. and R.A. 9h. 19m. 30s. Declination between 50° 34′ and 52° 15′ North.

Scale —1 millimetre to 24 seconds of arc.

Co-ordinates of the Fiducial stars marked with dots for the epoch A.D. 1900.

Star (.)	D.M. No. 1496—Zone + 51°	...	R.A.	9h. 14m. 8·0s.	...	Dec.	N. 51° 43·4′	...	Mag.	8·2
,, (··)	,, ,. 1500	,, 51°	...	,, 9h. 15m. 45·4s.	...	,,	51° 1·9′	...	,,	8·4
,, (∴)	,, ,, 1635	,, 50°	...	,, 9h. 17m. 56·3s.	...	,,	50° 42·5′	...	,,	7·5
,, (∷)	,, ,, 1389	,, 52°	...	,, 9h. 18m. 1·2s.	...	,,	52° 0·6′	...	,,	6·7

The photograph was taken with the 20-inch reflector on April 12th, 1893, between sidereal time 9h. 38m. and 13h. 11m., with an exposure of the plate during three and a half hours.

REFERENCES.

N.G.C. 2841. G.C. 1823. h 584. ♄ I. 205.

Sir J. Herschel, in the G.C., describes the nebula as very bright; large; very much extended in the direction 150° 8′; very suddenly much brighter in the middle; equal to a star of the 10th magnitude.

Lord Rosse (*Obs. of Neb. and Cl., p.* 75) states that, in general appearance, the nebula is very like the one in *Andromeda*; nucleus a little nearer the following edge than the centre; thinks there is a small part of detached nebulosity which seems to have a connection with the other.

The Photograph shows the nebula to have a dense stellar nucleus, which is surrounded by a zone of dense nebulosity, outside which is a zone with little, if any, nebulosity in it; and again, there is outside this another zone, or broad ring, of nebulosity. The nebula as a whole seems to be at a stage of development more perfectly symmetrical than the *Great Nebula* in *Andromeda*. The zones, or rings, of nebulosity approach the regularity of the rings of *Saturn*, but they are not so sharply defined at the margins. The nebula is elliptical, because it is viewed at an acute angle to our line of sight.

Plate 25.

NEBULA ♄ I 56-57 LEONIS.

NEBULA ♄ I 56-57 LEONIS.

Nebula ♅. I. 56, 57 Leonis.

R.A. 9h. 26m. 31s.; Dec. N. 21° 57′.

The photograph covers the region between R.A. 9h. 24m. 44s. and R.A. 9h. 28m. 20s. Declination between 21° 26′ and 22° 29′ North.

Scale—1 millimetre to 15 seconds of arc.

Co-ordinates of the Fiducial stars marked with dots for the epoch A.D. 1900.

Star ( )	D.M. No. 2100—Zone +22°	...	R.A. 9h. 24m. 46·4s.	...	Dec. N. 22° 16·4′	...	Mag. 7·0
„ (··) „	„ 2041 „ 21°	...	„ 9h. 25m. 8·6s.	...	„ 21° 28·5′	...	„ 8·5
„ (∴) „	„ 2045 „ 21°	...	„ 9h. 25m. 25·1s.	...	„ 21° 47·9′	...	„ 9·3
„ (∷) „	„ 2102 „ 22°	...	„ 9h. 26m. 34·2s.	...	„ 22° 17·9′	...	„ 7·2

The photograph was taken with the 20-inch reflector on April 4th, 1893, between sidereal time 9h. 5m. and 13h. 9m., with an exposure of the plate during four hours.

REFERENCES.

N.G.C. 2903 and 2905. G.C. 1861 and 1863. *h* 604, 1 and 604, 2. ♅ I. 56 and 57.

Sir J. Herschel, in the G.C., p. 79, describes the first of the nebulæ as considerably bright; very large; extended; gradually much brighter in the middle; barely resolvable; *south preceding* of 2. The second he describes as very faint; considerably large; round; pretty suddenly brighter in the middle; barely resolvable; *north following* of 2. A drawing of it is given in the *Phil. Trans.*, 1833, *pl. XV.*, *fig.* 70, but it does not resemble the photograph.

Lord Rosse (*Phil. Trans.*, 1850, *p.* 511, *pl. XXXVI.*, *fig.* 3) states that he saw the nebula as a spiral in an oblique direction resolved well, particularly towards the centre, where it is very bright. In the *Phil. Trans.*, 1861, *p.* 718, are given particulars of eighteen observations made between 1850 and 1858, and a marginal sketch indicates a spiral tendency. In the *Obs. of Neb. and Cl.*, *pp.* 76, 77, the results of many observations of the nebula are given, and on *pl. III.*, *fig.* 9, a drawing is given which shows several features resembling those shown on the photograph.

The Photograph clearly shows the nebula to be a spiral, having a well-defined stellar nucleus surrounded by nebulosity, and on the *north following* and *south preceding* sides are dense *ansæ* of nebulous matter, beyond and around which are two streams of dense nebulosity formed like the letter S, with the nucleus in the centre of figure. The two curved streams of nebulosity have star-like condensations involved in them at approximately equal distances from each other, and I can, on the negative, count five such condensations in each stream. The ends of the curved streams vanish gradually into invisibility, and can be traced about a diameter of the nebula beyond it.

The nebula *Messier* 81 *in Ursa Major* is very similar in character to that here described, and they forcibly suggest rotation round their respective nuclei, in the direction from *north following* to *south preceding*.

Plate 26.

NEBULÆ MESSIER 81 & 82 URSÆ MAJORIS.

Nebulæ Messier 81 and 82 Ursæ Majoris.

R.A. 9h. 47m. 18s. and 9h. 47m. 35s.; Dec. N. 69° 32·2′ and 70° 10·2′.

The photograph covers the region between R.A. 9h. 39m. 14s. and R.A. 9h. 54m. 39s. Declination between 68° 56′ and 70° 37′ North.

Scale—1 millimetre to 24 seconds of arc.

Co-ordinates of the Fiducial stars marked with dots for the epoch A.D. 1900.

Star (.) D.M. No. 584—Zone+70°	...	R.A. 9h. 44m. 53s.	...	Dec. N. 70° 20·7′	...	Mag. 8·1
,, (··) ,, ,, 542 ,, 69°	...	,, 9h. 46m. 51s.	...	,, 69° 22·4′	..	,, 8·5
,, (∵) ,, ,, 550 ,, 69°	...	,, 9h. 53m. 29s.	...	,, 69° 11·9′	...	,, 7·7

The photograph was taken with the 20-inch reflector on March 31st, 1889, between sidereal time 9h. 19m. and 12h. 49m., with an exposure of the plate during three and a half hours.

REFERENCES.

N.G.C. 3031, 3034. G.C. 1949=1953, 1950. h 649. ♅ IV. 79.

Sir J. Herschel, in the G.C., p. 81, describes the first of the nebulæ as remarkable; extremely bright; extended in the direction 156°; suddenly very much brighter in the middle, with a bright nucleus. The second nebula he describes as very bright; very large; very much extended (ray).

Lord Rosse (*Obs. of Neb. and Cl., pp.* 79, 80) describes M. 81 to be very like the nebula in *Andromeda,* and extending about 8′ from the nucleus to the north, but not beyond the two first stars, if so far; M. 82 he calls a most extraordinary object, at least 10 in length and crossed by several dark bands.

The Photograph shows the nebula, M. 81, to be a spiral, with a nucleus which is not well defined at its boundary, and is surrounded by rings of nebulous matter, but that the outermost rings are discontinuous in the *north preceding* and *south following* directions. It is very noticeable that there are numerous stars, or more probably star-like condensations, of the nebular matter arranged symmetrically, and apparently incorporated with the rings. The nebula extends far beyond the two stars referred to by Lord Rosse.

The nebula, M. 82, is probably seen edgewise with several nuclei of a nebulous character involved, and the rifts and attenuated places in it, which are clearly shown on the negative, are divisions between the rings, which would be visible as such if we could photograph the nebula from the direction perpendicular to its plane. We see it in section, and the upper and lower surfaces are very rugged.

Plate 27.

NEBULA ♄ I 168 URSÆ MAJORIS.

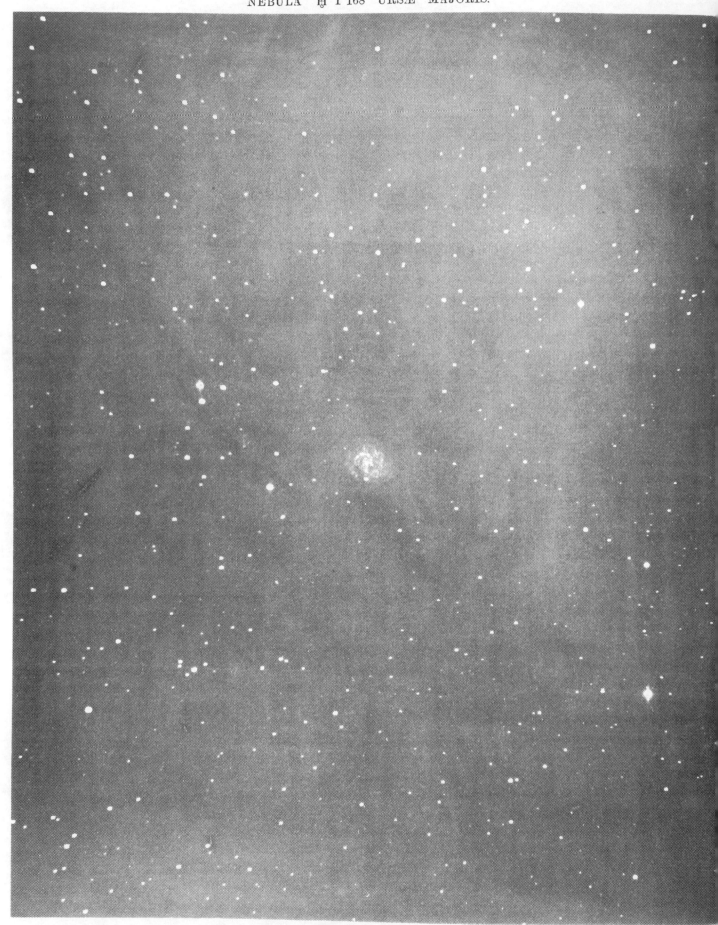

NEBULA ♄ I 168 URSÆ MAJORIS.

Nebula ♄. I. 168 Ursæ Majoris.

R.A. 10h. 12m. 15s.; Dec. N. 41° 55·1′.

The photograph covers the region between R.A. 10h. 8m. 32s. and R.A. 10h. 15m. 57s. Declination between 41° 4′ and 42° 45′ North.

Scale—1 millimetre to 24 seconds of arc.

Co-ordinates of the Fiducial stars marked with dots for the epoch A.D. 1900.

Star (.) D.M. No. 2104—Zone+42°	...	R.A. 10h. 9m. 23·6s.	...	Dec. N. 42° 21·7′	...	Mag. 7·5
„ (··) „ „ 2066 „ 41°	..	„ 10h. 10m. 39·0s. ...		„ 41° 46·4′	...	„ 7·3
„ (·⁚) „ „ 2072 „ 41°	...	„ 10h. 14m. 26·6s. ...		„ 41° 38·3′	....	„ 8·0
„ (⁚⁚) „ „ 2114 „ 42°	...	„ 10h. 15m. 6·5s. ...		„ 42° 21·0′	...	„ 6·7

The photograph was taken with the 20-inch reflector on April 14h, 1893, between sidereal time 9h. 54m. and 13h. 57m., with an exposure of the plate during four hours.

REFERENCES.

N.G.C. 3184.   G.C. 2052=2053.   h 688 = 689.   ♄ I. 168.

Sir J. Herschel, in the G.C., p. 83., describes two nebulæ in close proximity, one of them pretty bright; very large; round; very gradually brighter in the middle, and the other pretty faint.

Lord Rosse (*Obs. of Neb. and Cl., pp.* 82, 83, and *Phil. Trans.*, 1861, *p.* 719, *pl. XXVII., fig.* 13) delineates and describes the nebula with considerable reservation and doubt, but the drawing indicates that he saw part of a spiral structure with some very faint nebulosity.

The Photograph distinctly shows the nebula to be a spiral, with a well-defined nucleus equal to a star of about 15th magnitude in the centre of revolution, and around it, with geometrical precision, are the convolutions of the spiral.

It is remarkable in this, as in all other spiral nebulæ which I have photographed (some of which are included in this collection), that they have a star-like centre of revolution with the convolutions arranged symmetrically round it, and each volute is broken up into star-like loci at approximately equal distances from each other.

The records of these facts are now too numerous to be passed over as if they were the results of fortuitous circumstances. There must be a uniform cause to produce such uniform effects. What then is the cause? That which appears to me most probable is the collision of two nebulæ, or of two streams of meteoric matter, the bodies moving from opposite directions in space and meeting not quite centrally, but at some distance from the central line of motion. This determines the amplitude of the spiral and the form of the vortex, together with the position of the central nucleus.

Plate 28.

NEBULA MESSIER 94 CANUM VENATICORUM.

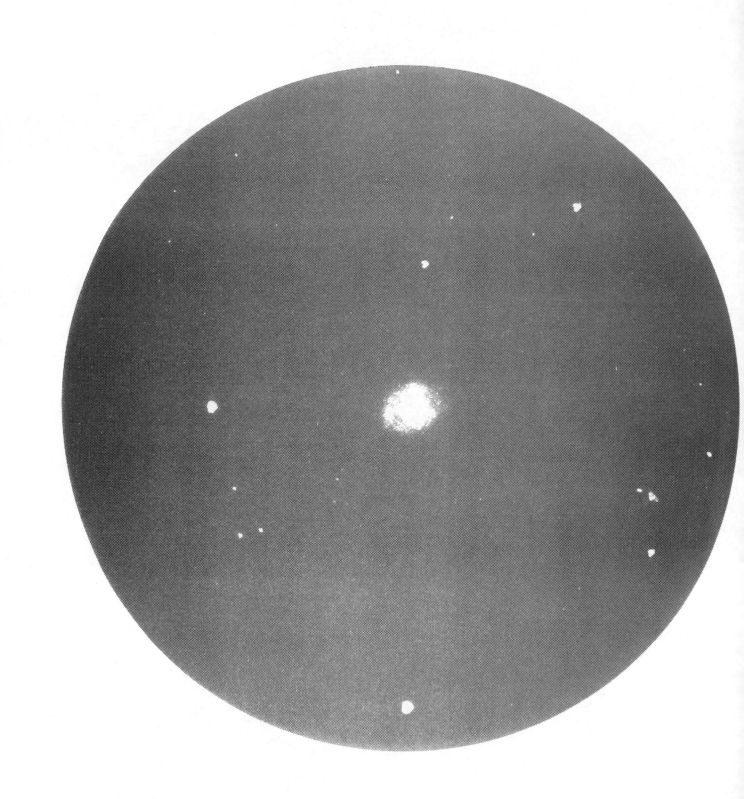

Nebula Messier 94 Canum Venaticorum.

The photograph covers a region 19 minutes of arc in diameter, with the nebula in the centre of the circle.

Scale—1 millimetre to 6 seconds of arc.

Co-ordinates of the centre of the nebula for the epoch A.D. 1900.

Nebula = D.M. No. 2333—Zone + 41° ... R.A. 12h. 46m. 13·1s. ... Dec. N. 41° 39·5′ ... Mag. 8·5

The photograph was taken with the 20-inch reflector on May 23rd, 1892, between sidereal time 14h. 24m. and 15h. 54m., with an exposure of the plate during ninety minutes.

REFERENCES.

N.G.C. 4736. G.C. 3258. *h* 1456. M. 94.

Sir J. Herschel, in the G.C., p. 105, describes the nebula as very bright; large; irregularly round; very suddenly much brighter in the middle; with bright nucleus; barely resolvable; and in the *Phil. Trans*, 1833, *p.* 434, *pl. XIII., fig.* 41, fuller details are given, but the drawing does not convey an idea of the nebula as it is shown on the photograph.

Lord Rosse (*Phil. Trans.*, 1861, *p.* 727, *and Obs. of Neb. and Cl., p.* 122) saw the nebula very little extended on the *preceding* and *following* sides; dark ring round the nucleus; then a bright ring exterior to this. The annulus, however, is not perfect, but broken up and patchy, and the object will probably turn out to be a spiral. There is much faint outlying nebulosity. These observations are extended in the *Obs. of Neb and Cl., p.* 122.

Lassell (*Mem. R.A.S., Vol. XXXVI., p.* 46, *pl. V., fig.* 25) describes the nebula as bright and round, increasing in brightness gradually to the centre, with a remarkable dark annulus. The figure illustrates this description.

The Photograph shows a large, distinct stellar nucleus in the midst of nebulosity, outside of which is a fainter zone, and outside that again is a denser zone like a broad ring, but not quite symmetrical in breadth. This outer ring is broken up into irregular star-like condensations, about ten of which can, on the negative, be counted. On the *south following* and *north preceding* sides are two abrupt projections from the ring, which probably led Lord Rosse to infer that the nebula has a spiral structure, but I am unable to trace any spiral structure on the photograph. Beyond the part of the nebula here described, there is a wide zone of very faint nebulosity symmetrically arranged round the dense central part, with its outer margin defined by a ring of slightly denser nebulosity. A longer exposure of the plate will be required to satisfactorily show this faint zone, and such long exposure will over-expose the central part, so that the details cannot be shown on a print from the negative, though they will be clearly visible to the eye through the nebulosity.

Plate 29.

CLUSTER MESSIER 53 COMÆ BERENICIS.

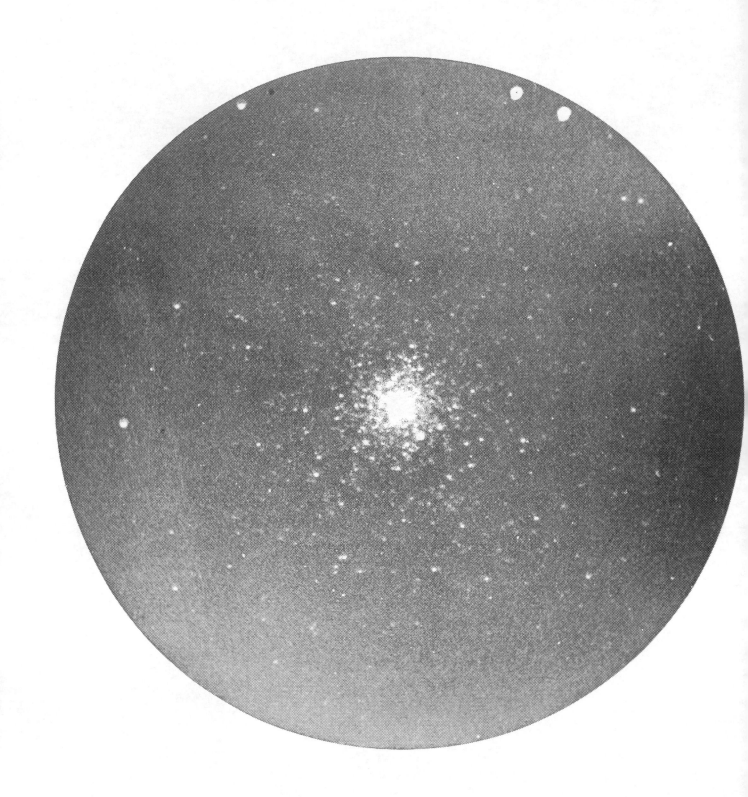

Plate 29.

Cluster Messier 53 Comæ Berenicis.

R.A. 13h. 8m. 2s.; Dec. N. 18° 41·9′.

Diameter of the field—19 minutes of arc.

Scale—1 millimetre to 6 seconds of arc.

Co-ordinates of the Fiducial star marked with a dot for the epoch A.D. 1900.

Star (.) D.M. No. 2702—Zone + 18° ... R.A. 13h. 8m. 15·2s. ... Dec. N. 18° 33·7′ ... Mag. 9·1

The photograph was taken with the 20-inch reflector on April 4th, 1892, between sidereal time 13h. 45m. and 14h. 45m., with an exposure of the plate during sixty minutes.

REFERENCES.

N.G.C. 5024. G.C. 3453. h 1558.

Sir J. Herschel, in the G.C., p. 108, describes it as a remarkable object; globular cluster; bright; very compressed; irregular and barely resolvable; very, very much brighter in the middle; stars 12th magnitude.

Lord Rosse (*Obs. of Neb. and Cl., p.* 125) names it a Globular Cluster, the more compressed part is about 75 seconds of arc in diameter, while the diameter of the whole cluster, excepting a few stragglers, is about 3 minutes in diameter.

The Photograph shows the cluster to consist of an aggregation of small stars close together, with the images of many of them overlapping, thus appearing like double and multiple stars. Like all other clusters of this class, which I have photographed, the central part is involved in nebulosity of a character sufficiently dense to obscure the stars on the print, though on the negative they are visible through it.

The component stars in the cluster have also the appearance of being nebulous, and are arranged in curves and lines with less dense nebulosity between them, so that they have the appearance of *lanes* leading in several directions.

Plate 30.

SPIRAL NEBULA MESSIER 51 CANUM VENATICORUM.

Spiral Nebula Messier 51 Canum Venaticorum.

R.A. 13h. 25m. 39s.; Dec. N. 47° 42·6′.

Diameter of the circle—19 minutes of arc.

Scale—1 millimetre to 6 seconds of arc.

Co-ordinates of the nebula for the epoch A.D. 1900.

R.A. 13h. 25m. 39s.; Dec. N. 47° 42·6′.

The photograph was taken with the 20-inch reflector on April 29th, 1889, between sidereal time 12h. 56m. and 16h. 59m., with an exposure of the plate during four hours.

REFERENCES.

N.G.C. 5194, 5195. G.C. 3572, 3574. *h* 1622, 1623. ⨳ I. 186.

Sir J. Herschel, in the G.C., p. 110, describes this as two nebulæ, the first, a magnificent object; with a nucleus and ring. The other bright; pretty small; round; very gradually brighter in the middle. In the *Phil. Trans.*, 1833, *pp.* 496, 497, *pl. X., fig.* 25, a detailed account is given, together with a drawing of the nebula, but it does not materially resemble the photograph.

Lord Rosse (*Phil. Trans.*, 1850, *pp.* 504, 505, *pl. XXXV., fig.* 1) describes the nebula, and shows in it a strong spiral structure which in several points resemble the photograph. In his *Obs. of Neb. and Cl., pp.* 127 *to* 131, *and pl. IV.*, the nebula is very fully described, and the drawing shows the general outline and some of the nebulous condensations in the spirals to be in good accord with the photograph. The observations and the drawings prove that immense labour has been expended in the production of such good work.

Lassell (*Mem. R.A.S., Vol. XXXVI., pp.* 46, 47, *pl. VI., figs.* 27, 27*a*) describes and delineates the nebula, and shows several parts of it in good accord with the photograph.

The Photograph shows both nuclei of the nebula to be stellar, surrounded by dense nebulosity, and the convolutions of the spiral in this as in other spiral nebulæ are broken up into star-like condensations with nebulosity around them. Those stars that do not conform to the trends of the spirals have nebulous trails attached to them, and seem as if they had broken away from the spirals. This appearance is conspicuous in two stars on the *preceding* side near the outer volute and in the notch on the *following* side of the middle volute. The notch on the *north following* side of the central nucleus is also due to the same cause. Hereafter it will be possible to follow the changes that may take place in this nebula by comparing this with other photographs.

Plate 31.

CLUSTER MESSIER 3 CANUM VENATICORUM.

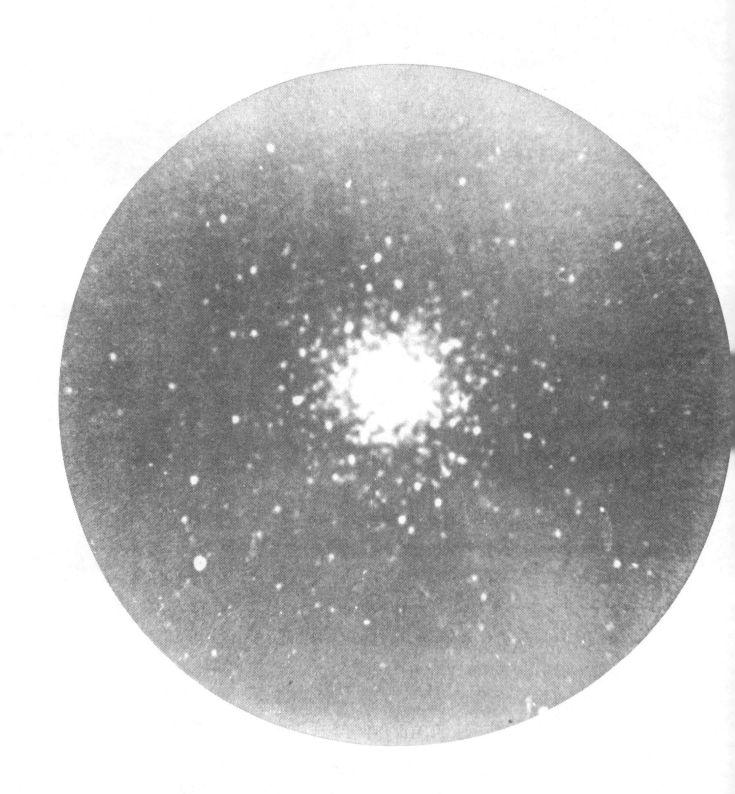

Cluster Messier 3 Canum Venaticorum.

R.A. 13h. 37m. 35s.; Dec. N. 28° 52·9′.

Diameter of the circle—19 minutes of arc.

Scale—1 millimetre to 6 seconds of arc.

Co-ordinates of the Fiducial star marked with a dot for the epoch A.D. 1900.

Star (.) D.M. No. 2448 —Zone + 29°   ...   R.A. 13h. 37m. 9·5s.   ...   Dec N. 28° 57·6′   ...   Mag. 9·1

The photograph was taken with the 20-inch reflector on May 2nd, 1891, between sidereal time 14h. 31m. and 16h. 31m., with an exposure of the plate during two hours.

REFERENCES.

N.G.C. 5272.   G.C. 3636.   h 1663.

Sir J. Herschel, in the G.C., p. 111., describes it as a very remarkable object; globular cluster; exceedingly bright; very large; very suddenly much brighter in the middle; stars 11th magnitude.

Lord Rosse (*Obs. of Neb. and Cl., p.* 132) describes the cluster as round and symmetrical; stars running into a blaze in the centre; rays running out on every side. Radiating branches somewhat resembling a St. Andrew's cross; central mass globular. There seemed to be a bifurcated dark lane in the north segment of the nucleus.

The Photograph confirms the general descriptions cited, though the print fails to show the stars that, on the negative, crowd the space covered by the dense nebulosity. The stars are clearly seen through the nebulosity on the negative, but the question obviously arises—Is the nebulosity due to the glare caused by the collective light of the mass of stars in the central part of the cluster?

The question pertains to all the globular clusters which I have photographed, and in its decision we require spectroscopic evidence which is not at present available, but I think the balance of evidence, such as we have, is in favour of it being real nebulosity which forms part of the cluster. And the fact that the two well-known clusters in the *sword-hand of Perseus*, also M. 11 Antinoi, and M. 37 Aurigæ, which are composed of close bright stars, do not show the least trace of nebulosity, goes to show that even bright masses of stars do not produce atmospheric glare resembling true nebulosity.

Plate 32.

NEBULA MESSIER 101 URSÆ MAJORIS.

NEBULA MESSIER 101 URSÆ MAJORIS.

Nebula Messier 101 Ursæ Majoris.

R.A. 13h. 59m. 39s.; Dec. N. 54° 49·8′.

The photograph covers the region between R.A. 13h. 55m. 2s. and R.A. 14h. 4m. 22s. Declination between 54° 9′ and 55° 50′ North.

Scale—1 millimetre to 24 seconds of arc.

Co-ordinates of the Fiducial stars marked with dots for the epoch A.D. 1900.

Star (.) D.M. No. 1640—Zone + 54°	...	R.A. 13h. 56m. 57·5s.	..	Dec. N. 54° 14·2′	...	Mag. 8·7
„ (··) „ „ 1649 „ 55°	...	„ 13h. 57m. 50·4s.	...	„ 54° 50·4′	...	„ 8·5
„ (∴) „ „ 1650 „ 55°	...	„ 13h. 59m. 18·7s.	...	„ 55° 8·8′	...	„ 7·8
„ (∷) „ „ 1660 „ 55°	...	„ 14h. 2m. 3·7s.	...	„ 54° 47·2′	...	„ 9·4

The photograph was taken with the 20-inch reflector on May 30th, 1892, between sidereal time 14h. 32m. and 17h. 55m., with an exposure of the plate during three hours and twenty minutes.

REFERENCES.

N.G.C. 5457, 5458.   G.C. 3770, 3771.   h 1744.

Sir J. Herschel, in the G.C., p. 114, describes the nebula as pretty bright; very large; irregularly round; very suddenly much brighter in the middle, with a bright small nucleus; and connected with it is another nebula very faint; pretty large; round; very little brighter in the middle.

Lord Rosse (*Phil. Trans.*, 1861, *p.* 729, *pl. XXIX.*, *fig.* 35) describes the nebula as a large spiral; faintish; several arms and knots; 14′ across at least; has a nucleus; light very patchy. He also gives a rough sketch as well as the drawing of it, which in outline are in very fair accord with the photograph. In the *Obs. of Neb. and Cl*, *p.* 135, are given further observations confirmatory of the foregoing, and a sketch is also given.

The Photograph shows the nebula to be a spiral with a well-defined stellar nucleus, and each of the convolutions is broken up into numerous star-like condensations. The outer extensions of the nebulosity can be traced on the negative farther than on the print, and they show the structure and details for the first time as they actually exist.

On the *north preceding* and *north following* sides of the nebula are shown the five nebulæ following, and their numbers in the *New General Catalogue* are: 5422 ♄.I. 230. 5473 ♄.I. 231.   5477 ♄.III. 790.   5485 ♄.I. 232.   5486 ♄.II. 801.

M

Plate 33.

CLUSTER MESSIER 5 LIBRÆ.

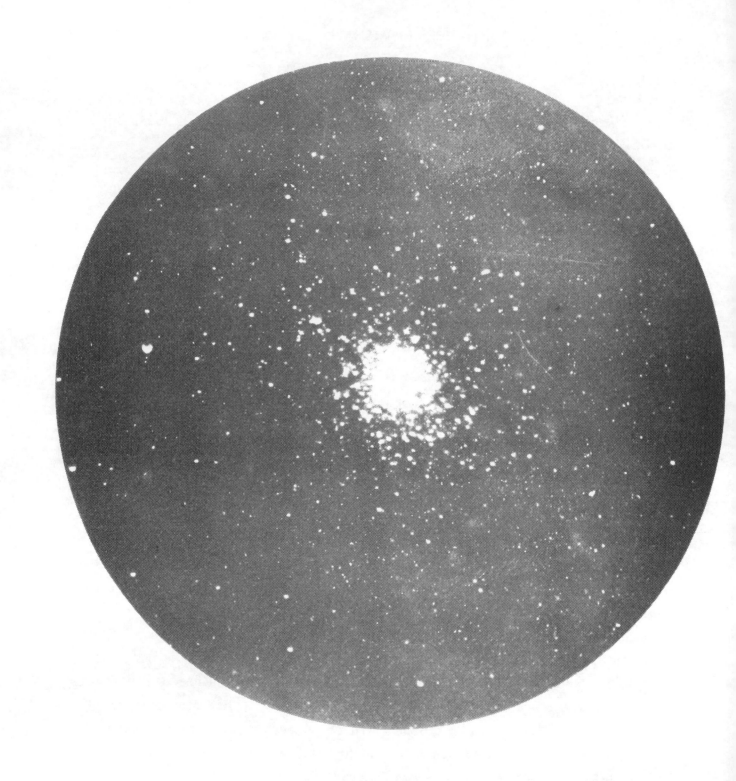

Cluster Messier 5 Libræ.

Diameter of the circle —19 minutes of arc.

Scale—1 millimetre to 6 seconds of arc.

Co-ordinates of the centre of the cluster for the epoch A.D. 1900.

R.A. 15h. 13m. 29s. ; Dec. N. 2° 26·8′.

The photograph was taken with the 20-inch reflector on April 25th, 1892, between sidereal time 14h. 49m. and 15h. 49m., with an exposure of the plate during sixty minutes.

REFERENCES.

N.G.C. 5904. G.C. 4083. *h* 1916.

Sir J. Herschel (*Phil. Trans.*, 1833, *p.* 456, *pl. XVI., fig.* 87) describes it as a most magnificent, excessively compressed cluster of a globular character. Stars 11th to 15th magnitude, and the condensation is progressive up to the centre, where the stars run together into a blaze or like a snowball. A further reference is also given in the G.C., p. 119.

The drawing only indicates the character of the cluster, and therefore cannot be compared with the photograph.

Lord Rosse (*Obs. of Neb. and Cl., p.* 146) states that the cluster is more than 7 or 8 in diameter, with a nebulous appearance in the centre (where the stars are very dense), and in spirals among the stars.

The Photograph shows the stars to about 15th magnitude, and the cluster is involved in dense nebulosity about the centre. The nebulosity hides the stars even on the negative.

Plate 34.

CLUSTER MESSIER 13 HERCULIS.

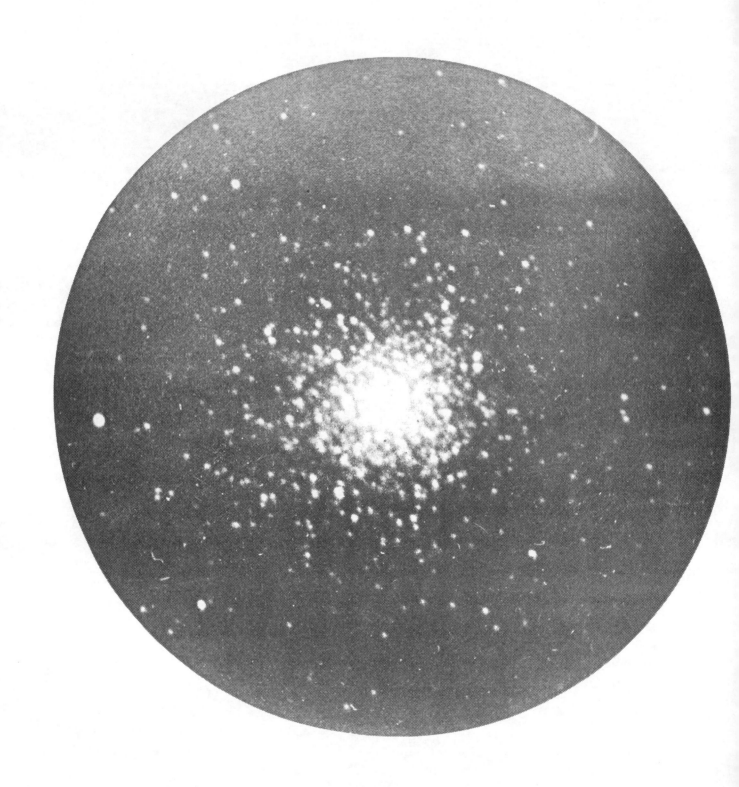

Plate 34.

Cluster Messier 13 Herculis.

Diameter of the circle—19 minutes of arc.

Scale—1 millimetre to 6 seconds of arc.

Co-ordinates of the centre of the cluster for the epoch A.D. 1900.

R.A. 16h. 38m. 6s.; Dec. N. 36° 39·0′.

The photograph was taken with the 20-inch reflector on May 22nd, 1887, between sidereal time 16h. 19m. and 17h. 19m., with an exposure of the plate during sixty minutes.

REFERENCES.

N.G.C. 6205. G.C. 4230. *h* 1968.

Sir J. Herschel (*Phil. Trans.*, 1833, *p.* 458, *pl. XVI.*, *fig.* 86) describes it as a very rich cluster; irregularly round; very large; very gradually much brighter in the middle; stars 10th to 15th magnitude, of which there must be thousands; does not come up to a nucleus; has hairy-looking curvilinear branches. The drawing (*pl. XVI.*, *fig.* 86) gives an indication of the general character of the cluster, but as the stars are not laid down to scale, it cannot be compared with the photograph.

Lord Rosse (*Phil. Trans.*, 1861, *p.* 732, *pl. XXVIII.*, *fig.* 33, *and Obs. of Neb. and Cl.*, *p.* 150) states that the cluster seems to have a dark streak across the bright part, a little above the centre. Dark spaces seen through mist; dark lanes.

The Photograph clearly shows the dark lanes referred to by Lord Rosse, and that they meet in the cluster, forming a figure like the letter Y. There are also other lanes of somewhat similar character, in and around the cluster, which are visible on the negative; and the mist referred to is the nebulosity that seems invariably to be present in globular clusters. The stars also seem to be themselves nebulous, and are clearly seen, on the negative, through the nebulosity.

Plate 35.

CLUSTER MESSIER 12 OPHIUCHI

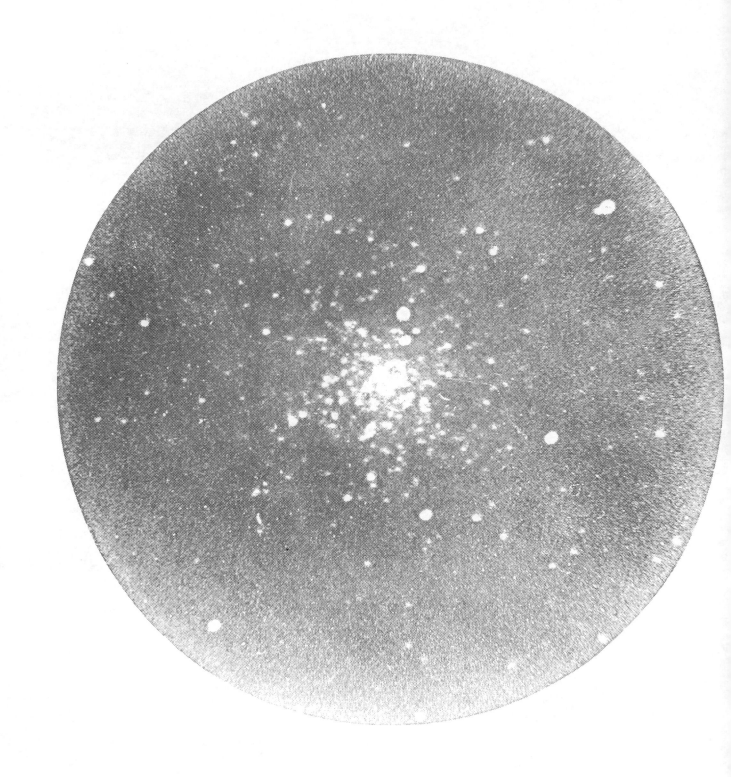

Cluster Messier 12 Ophiuchi.

R.A. 16h, 42m. 2s.; Dec. S. 1° 46·2′.

Diameter of the circle—19 minutes of arc.

Scale—1 millimetre to 6 seconds of arc.

Co-ordinates of the Fiducial stars marked with dots for the epoch A.D. 1900.

Star (.)·D.M. No. 3246—Zone—1° ... R.A. 16h. 42m. 22·2s. ... Dec. S. 1° 44·6′ ... Mag. 9·5

„ (··) „ „· 3247 „ 1° ... „ 16h. 42m. 27·8s. ... „ 1° 51·9′ ... „ 9·5

The photograph was taken with the 20-inch reflector on June 27th, 1892, between sidereal time 16h. 39m. and 18h. 39m., with an exposure of the plate during two hours.

REFERENCES.

N.G.C. 6218.   G.C. 4238.   h 1971.

Sir J. Herschel, in the G.C., p. 122, describes it as a very remarkable globular cluster; very bright; very large; irregularly round; gradually much brighter in the middle; well resolved; stars 10th magnitude.

Lord Rosse (*Phil. Trans.*, 1861, *p. 732, and Obs. of Neb. and Cl., p.* 150) states that the branches of the cluster have a slightly spiral arrangement, dark lanes, but not remarkable.   Centre quite nebulous with power 414; the stars there cannot be brighter than 17th magnitude.

The Photograph shows the stars to be visible from the outer limits to the centre, and though there is a trace of nebulosity to be seen, it is much less dense than that visible in other globular clusters.   The dark lanes referred to are conspicuously visible, but a spiral arrangement is not remarkably striking.

Plate 36.

CLUSTER MESSIER 10 OPHIUCHI.

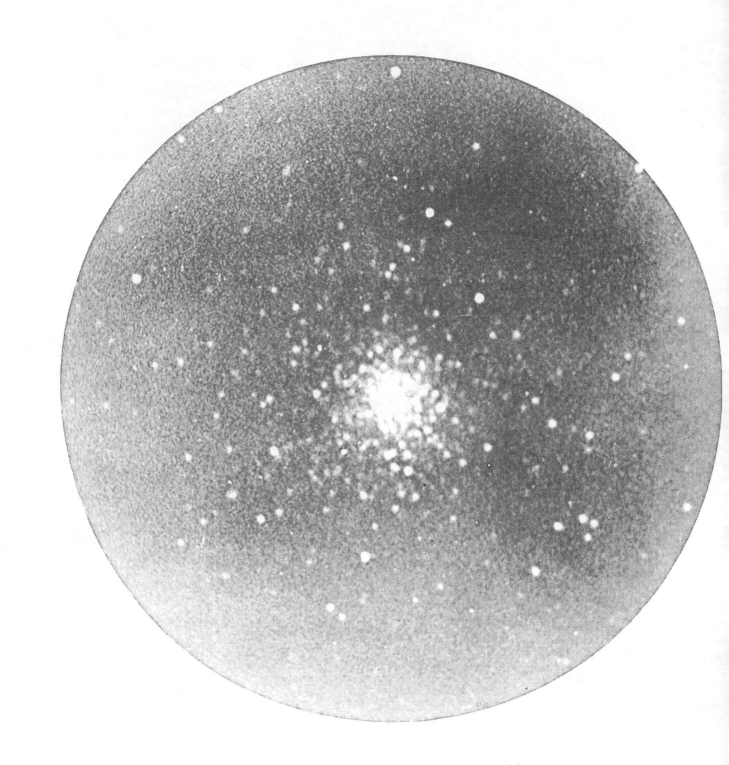

Plate 36.

Cluster Messier 10 Ophiuchi.

Diameter of the circle—19 minutes of arc.

Scale—1 millimetre to 6 seconds of arc.

Co-ordinates of the centre of the cluster for the epoch A.D. 1900.

R.A. 16h. 51m. 53s.; Dec. S. 3° 56·7′.

The photograph was taken with the 20-inch reflector on June 9th, 1891, between sidereal time 16h. 9m. and 17h. 53m., with an exposure of the plate during seventy-five minutes.

REFERENCES.

N.G.C. 6254.   G.C. 4256.   h 1972=3659.

Sir J. Herschel, in the G.C., p. 122, states that it is a remarkable globular cluster; bright; very large; round; gradually very much brighter in the middle; well resolved; stars 10th to 15th magnitude.

Lord Rosse (*Phil. Trans.*, 1861, *p. 732, and Obs. of Neb. and Cl., p.* 150) states that at the north end is a series of bright stars forming a segment of a circle with the convexity south.   A dark lane (not very conspicuous) above the middle and quite across the cluster.

The Photograph shows several curves of stars which are visible to the centre of the cluster, and that it is nearly free from nebulosity, so that the stars are not much obscured by it.   No doubt more nebulosity would be shown with a longer exposure of the plate than during 75 minutes, and the star images would be brighter and larger.

Plate 37.

CLUSTER MESSIER 92 HERCULIS.

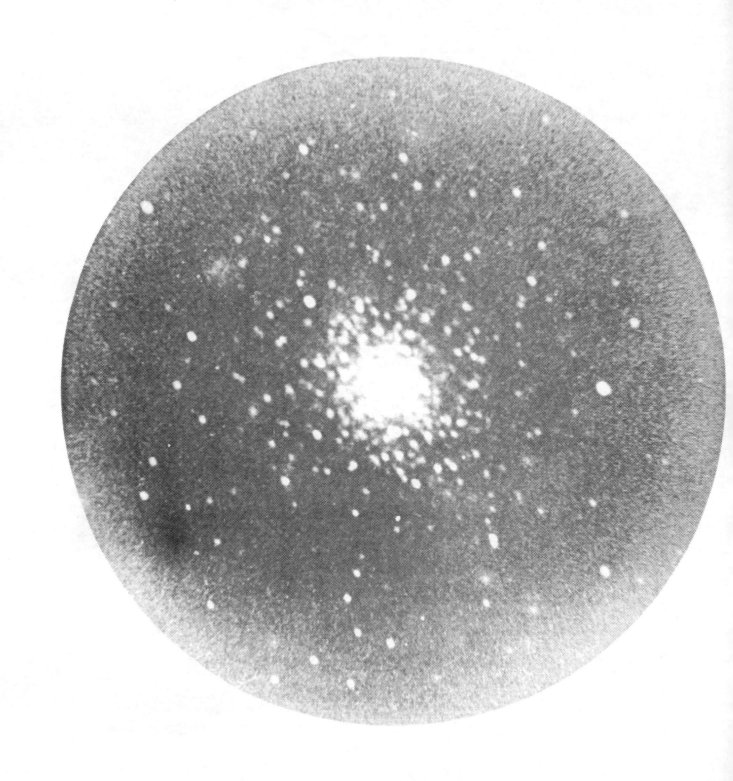

Cluster Messier 92 Herculis.

R.A. 17h. 14m. 5s.; Dec. N. 43° 14·7′.

Diameter of the circle—19 minutes of arc.

Scale—1 millimetre to 6 seconds of arc.

Co-ordinates of the Fiducial star marked with a dot for the epoch A.D. 1900.

Star (.) D.M. No. 2712—Zone+43° ... R.A. 17h. 14m. 36·9s. ... Dec. N. 43° 15·3′ ... Mag. 9·0

The photograph was taken with the 20-inch reflector on May 31st, 1891, between sidereal time 16h. 55m. and 17h. 55m., with an exposure of the plate during sixty minutes.

REFERENCES.

N.G.C. 6341. G.C. 4294.

Sir J. Herschel, in the G.C., p. 123, describes it as a globular cluster; very bright; very large; exceedingly compressed in the middle; well resolved; stars small.

Lord Rosse (*Obs. of Neb. and Cl., p.* 151) states that the nucleus is possibly spiral; darker spaces. Nucleus barely, if at all, resolved.

The Photograph shows the cluster to be involved in dense nebulosity, which on the negative almost prevents the stars being seen through it, and on the print quite obscures the stars. The stars in this, as in all other globular clusters, are arranged in various patterns, and many of them appear to be nebulous.

Plate 38.

NEBULA MESSIER 17 CLYPEI.

NEBULA MESSIER 17 CLYPEI.

Nebula Messier 17 Clypei.

R.A. 18h. 15m. 0s.; Dec. S. 16° 13·1.

The photograph covers the region between R.A. 18h. 12m. 16s. and R.A. 18h. 17m. 55s. Declination between 15° 23′ and 17° 4′ South.

Scale—1 millimetre to 24 seconds of arc.

Co-ordinates of the Fiducial stars marked with dots for the epoch A.D. 1900.

Star	D.M. No.	Zone	R.A.	Dec. S.	Mag.
(.)	No. 4806	—16°	18h. 13m. 17·4s.	16° 19·8′	8·8
(··)	„ 4805	„ 16°	„ 18h. 13m. 16·8s.	„ 16° 41·8′	„ 8·5
(∵)	„ 4927	„ 15°	„ 18h. 14m. 23·4s.	„ 15° 52·5′	„ 5·7
(∷)	„ 4836	„ 16°	„ 18h. 16m. 2·4s.	„ 16° 22·0′	„ 7·8

The photograph was taken with the 20-inch reflector on August 5th, 1893, between sidereal time 18h. 25m. and 20h. 25m., with an exposure of the plate during two hours.

REFERENCES.

N.G.C. 6618. G.C. 4403. *h* 2008.

Sir J. Herschel (*Phil. Trans.*, 1833, *p.* 461, *pl. XII., fig.* 35, *and G.C., p.* 125) describes it as a magnificent object; bright; extremely large; extremely irregular figure, two-hooked. The drawing conveys an idea of the brighter portions of the nebula as shown on the photograph.

Lord Rosse (*Obs. of Neb. and Cl., p.* 151, *pl. VI., figs.* 1, 2) describes and depicts the nebula in a manner much more resembling the photograph in the brighter parts, but, of course, the details cannot be compared.

Lassell (*Mem. R.A.S., Vol. XXXVI., p.* 49, *pls. VII. and VIII., figs.* 33, 33*a*) describes and delineates the nebula (more than the observers referred to above) in harmony with the photograph.

The Photograph shows clearly the bright nebulosity which Herschel, Rosse, and Lassell describe, together with the true relative forms and density of the several patches of the nebulous condensations and of the stars involved. It moreover shows, very faintly, extensions of the nebulosity surrounding the bright parts, completing the figure of the nebula into an oval, with the long, bright limb of nebulosity forming the semi-axis major connecting the centre with the *north preceding* end where the " *Omega* " part is seen. The major axis of the nebula lies *south following* to *north preceding,* and measures about 18 minutes of arc in length; the minor axis measures about 12 minutes.

The nebula is too far south in this latitude to obtain the best photographic results, but those here given are far in advance of the descriptions hitherto published, and the chart shows thousands of very interesting stars, star-groups, and apparent double, triple, and multiple stars, that will provide a large field for investigation in the future.

Plate 39.

CLUSTER MESSIER 26 CLYPEI.

Cluster Messier 26 Clypei.

R.A. 18h. 39m. 45s. ; Dec. S. 9° 29·8′.

The photograph covers the region between R.A. 18h. 36m. 57s. and R.A. 18h. 42m. 28s. Declination between 8° 39′ and 10° 20′ South.

Scale—1 millimetre to 24 seconds of arc.

Co-ordinates of the Fiducial stars marked with dots for the epoch A.D. 1900.

Star (.) D.M. No. 4811—Zone—9°	..	R.A. 18h. 38m. 23·8s.	...	Dec. S. 9° 1·9′	...	Mag. 8·3
„ (··) „ „ 4815 „ 9°	...	„ 18h. 39m. 2·7s.	...	„ 9° 24·5′	...	„ 9·0
„ (∴) „ „ 4828 „ 9°	...	„ 18h. 40m. 42·5s.	...	„ 9° 34·4′	...	„ 9 0
„ (∷) „ „ 4832 „ 9°	...	„ 18h. 41m. 17·0s.	...	„ 9° 51·2′	...	„ 9·1

The photograph was taken with the 20-inch reflector on August 15th, 1892, between sidereal time 18h. 44m. and 20h. 15m., with an exposure of the plate during ninety minutes.

REFERENCES.

N.G.C. 6694.   G.C. 4432.   h 3758.

Sir J. Herschel, in the G.C., p. 126, describes the cluster as considerably large; pretty rich; pretty compressed; stars 12th to 15th magnitude.

The Photograph shows the cluster and the region surrounding, in which it will be observed that the stars are pretty numerous and small in magnitude. There are among them several combinations of star-groups, double, triple, and apparently multiple stars.

Plate 40.

CLUSTER   MESSIER 11   ANTINOI.

CLUSTER   MESSIER 11   ANTINOI.

Cluster Messier 11 Antinoi.

R.A. 18h. 45m. 42s.; Dec. S. 6° 23·3′.

The photograph covers the region between R.A. 18h. 44m. 2s. and R.A. 18h. 47m. 25s. Declination between 5° 50′ and 6° 53′ South.

Scale—1 millimetre to 15 seconds of arc.

Co-ordinates of the Fiducial stars marked with dots for the epoch A.D. 1900.

Star (.) D.M. No. 4922—Zone—6°	...	R.A. 18h. 44m. 19·8s.	...	Dec. S. 6° 1·3′	...	Mag. 6·8
„ (··) „ „ 4924 „ 6°	...	„ 18h. 44m. 41·4s.	...	„ 6° 36·7′	...	„ 9·1
„ (∵) „ „ 4935 „ 6°	...	„ 18h. 46m. 10·8s.	...	„ 6° 4·8′	...	„ 9·6
„ (∷) „ „ 4936 „ 6°	...	„ 18h. 46m. 16·7s.	...	„ 6° 24·7′	...	„ 9·5

The photograph was taken with the 20-inch reflector on September 22nd, 1891, between sidereal time 19h. 56m. and 20h. 51m., with an exposure of the plate during fifty-five minutes.

REFERENCES.

N.G.C. 6705. G.C. 4437. *h* 2019.

Sir J. Herschel, in the G.C., p. 126, describes it as a remarkable cluster; very bright; large; irregularly round; rich; stars 9th to 11th magnitude.

Lord Rosse (*Obs. of Neb. and Cl., p.* 151) describes it as a cluster curiously broken up into groups, one star two or three classes brighter than the rest.

The Photograph shows the cluster consistently with the description by Lord Rosse, and the negative shows the stars individually, though the print, owing to their closeness, does not separate them. It will be observed that there are many apparent double, triple, and multiple stars in the cluster and in the surrounding region, and it is entirely free from nebulosity.

Plate 41.

"RING" NEBULA MESSIER 57 LYRÆ.

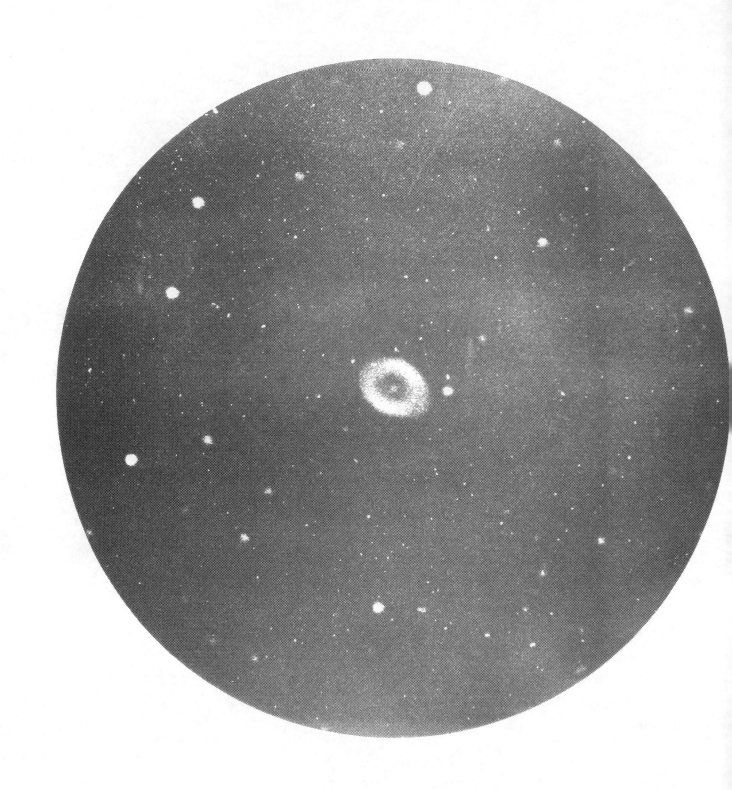

Plate 41.

"Ring" Nebula Messier 57 Lyræ.

Diameter of the field—12 minutes 40 seconds of arc.

Scale—1 millimetre to 4 seconds of arc.

Co-ordinates of the centre of the nebula for the epoch A.D. 1900.

R.A. 18h. 49m. 52·5s.; Dec. N. 32° 54·0′ (=D.M. 3246+32°, 9·0′ mag.).

The photograph was taken with the 20-inch reflector on July 27th, 1891, between sidereal time 20h. 32m. and 21h. 2m., with an exposure of the plate during thirty minutes.

REFERENCES.

N.G.C. 6720.   G.C. 4447.   *h* 2023.

Sir J. Herschel (*Phil. Trans.*, 1833, *p.* 462, *pl. X., fig.* 29, *and G.C., p.* 126) describes the nebula and the small star on the *following* side, which is exactly on the parallel of the centre, and distance equal to the breadth of the ring. The central vacuity of the ring is not black; a nebulous light fills it. The drawing agrees with these descriptions.

Lord Rosse (*Phil. Trans.*, 1844, *p.* 322, *pl. XIX., fig.* 29, *and Obs. of Neb. and Cl., p.* 152) delineates and describes the nebula in considerable detail, and shows filaments projecting from the edge.

The Photograph shows the nebula and the interior of the ring more elliptical than the drawings and the descriptions indicate; and the star on the *following* side is nearer to the ring than the distance given. The nebulosity on the *preceding* and *following* ends of the ring protrudes a little and is less dense than on the *north* and *south* sides.

This probably suggested the filamentous appearance which Lord Rosse shows. Seven photographs of the nebula have been taken between 1887 and 1891, and the central star is strongly shown on some of them, but on others it is scarcely visible, which points to the star being variable.

Plate 42.

REGION OF THE CLUSTER MESSIER 56 LYRÆ.

REGION OF THE CLUSTER MESSIER 56 LYRÆ.

Region of the Cluster Messier 56 Lyræ.

R.A. 19h. 12m. 42s.; Dec. N. 30° 0·4′.

The photograph covers the region between R.A. 19h. 9m. 28s. and R.A. 19h. 15m. 41s. Declination between 29° 10′ and 30° 51′ North.

Scale—1 millimetre to 24 seconds of arc

Co-ordinates of the Fiducial stars marked with dots for the epoch A.D. 1900.

Star (.) D.M. No. 3479—Zone +30° ... R.A. 19h. 10m. 27·4s. ...	Dec. N. 30° 10·7′ · ...	Mag. 7·8
„ (··) „ „ 3527 „ 29° ... „ 19h. 10m. 38·1s. ...	„ 29° 34·3′ ...	„ 8·6
„ (∵) „ „ 3491 „ 30° ... „ 19h. 11m. 33·6s. ...	„ 30° 21·1′ ...	„ 5·8
„ (∷) „ „ 3550 „ 29° ... „ 19h. 14m. 32·1s. ...	„ 29° 46·3′ ...	„ 8·0

The photograph was taken with the 20-inch reflector on August 28th, 1892, between sidereal time 19h. 18m. and 22h. 21m., with an exposure of the plate during three hours.

REFERENCES.

N.G.C. 6779. G.C. 4485. *h* 2036.

Sir J. Herschel, in the G.C., p. 127, describes the cluster as globular; bright; large; irregularly round; gradually very much compressed in the middle; well resolved; stars 11th to 14th magnitude.

Lord Rosse (*Obs. of Neb. and Cl., p.* 152) states that a few scattered stars have the appearance of rays from the centre.

The Photograph shows the cluster with rays projecting from it, as suggested by Lord Rosse, and nebulosity in the centre. On the negative the stars are visible through the nebulosity to the centre of the cluster, and the region around is very richly covered with faint stars that will provide a large field for future investigations.

Plate 43.

CHART OF STARS IN CYGNUS.

Chart of Stars in Cygnus.

The photograph covers the region between R.A. 19h. 41m. 14s. and R.A. 19h. 50m. 26s. Declination between 34° 28′ and 36° 38′ North.

Scale—1 millimetre to 31 7 seconds of arc, equal to MM. Henry's scale.

Co-ordinates of the Fiducial stars marked with dots for the epoch A.D. 1900.

Star (.) D.M. No. 3786—Zone + 35°	...	R.A. 19h. 41m. 59·9s.	...	Dec. N. 35° 51·0′	...	Mag. 7·0
„ (··) „ „ 3701 „ 34°	...	„ 19h. 42m. 7·4s.	...	„ 34° 45·9′	...	„ 6·6
„ (·:) „ „ 3727 „ 34°	..	„ 19h. 45m. 0·2s.	...	„ 35° 3·4′	...	„ 7·0
„ (::) „ „ 3826 „ 35°	...	„ 19h. 46m. 39·2s.	..	„ 35° 50·9′	...	„ 7·5
., (:·) „ „ 3744 „ 36°	...	„ 19h. 48m. 37·1s.	...	„ 36° 10·3′	...	„ 6·3

The photograph was taken with the 20-inch reflector on August 14th, 1887, between sidereal time 19h. 29m. and 20h. 29m., with an exposure of the plate during one hour.

A photograph of this region was taken by MM. Henry, at the Paris Observatory, on August 14th, 1885, with the 13¼-inch refractor constructed by them for charting the stars. This was adopted in 1887 as the model for the international refractors, which now number 18, engaged in the charting.

This Photograph exhibits one of the difficulties connected with charting the stars by photography, which will be appreciated by the following illustration :—On August 14th, 1858, MM. Henry took a photograph with the 13¼-inch refractor of the region of the sky coincident with that on the plate annexed, and I counted upon it 3124 stars. On August 23rd, 1886, I took, with the 20-inch reflector, a photograph of the same region, and counted upon it 5023 stars. On August 14th, 1887, I took another photograph of this region, and counted upon it 16,200 stars. Each of the three photographs was exposed to the sky during one hour, and yet we find these enormous differences between them.

Plate 44.

"DUMB-BELL" NEBULA IN VULPECULA.

"DUMB-BELL" NEBULA IN VULPECULA.

" Dumb-bell " Nebula in Vulpecula.

R.A. 19h. 55m. 17s.; Dec. N. 22° 26·8'.

The photograph covers the region between R.A. 19h. 53m. 28s. and R.A. 19h. 57m. 8s. Declination between 21° 56' and 22° 59' North.

Scale—1 millimetre to 15 seconds of arc.

Co-ordinates of the Fiducial stars marked with dots for the epoch A.D. 1900.

Star (.)	D.M. No. 3871—Zone +22°	...	R.A. 19h. 54m. 44·5s. ...	Dec. N. 22° 21·2'	...	Mag. 9·2
„ (··)	„ „ 3874 „ 22°	...	„ 19h. 54m. 57·1s. ...	„ 22° 43·2'	...	„ 8·6
„ (·:)	„ „ 3884 „ 22°	...	„ 19h. 56m. 0·7s. ...	„ 22° 9·8'	...	„ 8·0
„ (::)	„ „ 3889 „ 22°	...	„ 19h. 56m. 37·5s. ...	„ 22° 34·3'	...	„ 8·1

The photograph was taken with the 20-inch reflector on October 3rd, 1888, between sidereal time 20h. 12m. and 23h. 20m., with an exposure of the plate during three hours.

REFERENCES.

N.G.C. 6853.   G.C. 4532.   *h* 2060.

Sir J. Herschel (*Phil. Trans.*, 1833, *p.* 497, *pl. X., fig.* 26) describes the nebula with considerable minuteness, and the drawing conveys a fair idea of its general outlines.

Lord Rosse (*Phil. Trans.*, 1844, *p.* 322, *pl. XIX., fig.* 26) describes and delineates the nebula, but the drawing does not compare well with the photograph, nor does that in the *Phil. Trans.*, 1850, *p.* 508, *pl. XXXVIII., fig.* 17.   In the *Phil. Trans.*, 1861, *p.* 705, *pl. XXXI., fig.* 43, further descriptions are given, and the drawing in its outline is a fair representation of the nebula.   The descriptive matter is supplemented on pp. 728, 729, with observations by Otto Struve.   In the *Obs. of Neb. and Cl., pp.* 154, 155, are given further particulars of the nebula and its contained stars.

Lassell (*Mem. R.A.S., Vol. XXXVI., p.* 50, *pl. IX., fig.* 35) also describes and delineates the nebula.

The Photograph shows more of the character and constitution of the nebula and the contained stars, or star-like condensations within it, than all the drawings and descriptive matter which have been presented to us.   The knowledge is obtained with very little effort on the part of the observer, and it has also the merit of being reliable.

The nebula is probably a globular mass of nebulous matter which is undergoing the process of condensation into stars, and the faint protrusions of nebulosity on the *south following* and *north preceding* ends are the projections of a broad ring of nebulosity which surrounds the globular mass.   This ring, not being sufficiently dense to obscure the light of the central region of the globular mass, is dense enough to obscure those parts of it that are hidden by the increased thickness of the nebulosity, thus producing the " dumb-bell " appearance.

If these inferences are true, we may proceed yet a step, or a series of steps, further, and predict that the consummation of the life history of this nebula will be its reduction to a globular cluster of stars.

Plate 45.

NEBULA IḤ V 15 CYGNI.

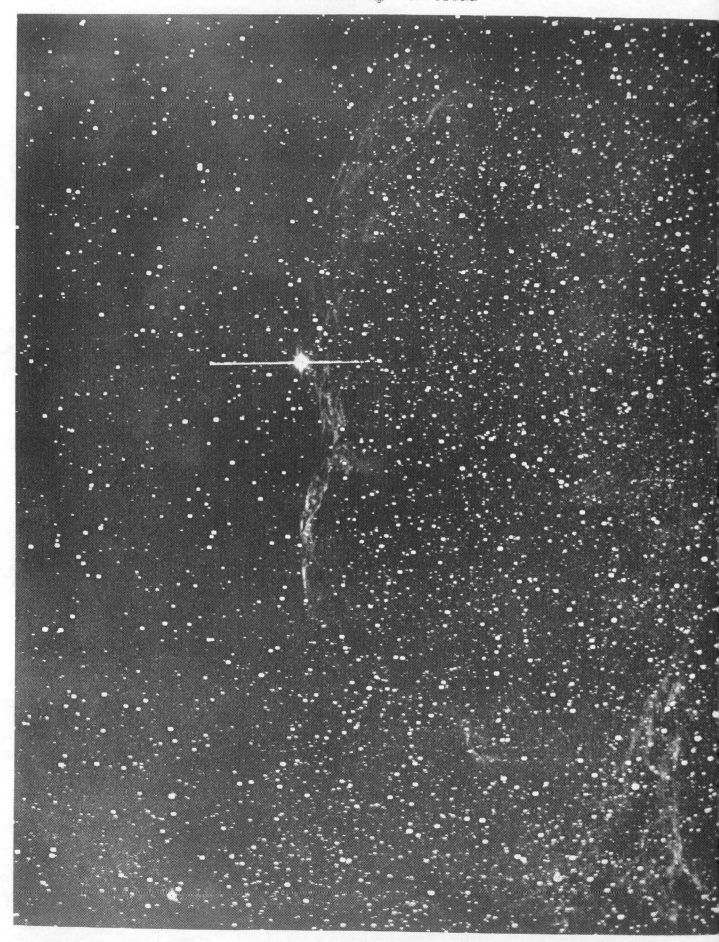

Nebula ♄. V. 15 Cygni.

R.A. 20h. 41m. 32s.; Dec. N. 30° 21·4′.

The photograph covers the region between R.A. 20h. 38m. 54s. and R.A. 20h. 45m. 14s. Declination between 29° 43′ and 31° 24′ North.

Scale—1 millimetre to 24 seconds of arc.

Co-ordinates of the Fiducial stars marked with dots for the epoch A.D. 1900.

Star (.) D.M. No. 4204—Zone +31°	...	R.A. 20h. 40m. 21·3s.	...	Dec. N. 31° 20·0′	...	Mag. 8·0
„ (··) „ „ 4167 „ 30°	...	„ 20h. 41m. 31·4s.	...	„ 30° 21·0′	...	„ 4·3
„ (∵) „ „ 4222 „ 31°	...	„ 23h. 43m. 41·7s.	...	„ 31° 18·0′	...	„ 7·8
„ (∷) „ „ 4185 „ 30°	...	„ 20h. 44m. 29·5s.	...	„ 30° 25·1′	...	„ 7·5

The photograph was taken with the 20-inch reflector on September 28th, 1891, between sidereal time 20h. 36m. and 0h. 40m., with an exposure of the plate during four hours.

REFERENCES.

N.G.C. 6960.   G.C. 4600.   h 2088.

Sir J. Herschel (*Phil. Trans.*, 1833, *p.* 498, *pl. XI., fig.* 33) thus describes the nebula:—" Very long and winding, and runs northward from *k* full two fields breadth (30′). The nebula extends southward far beyond *k* Cygni, but is extremely faint; the northern part is pretty bright, and extends to two stars. Northwards from *k* Cygni, 27′, extends a curved tail of nebula of a serpentine form, fading very gradually into two tails forming a fork." The drawing in its outline very fairly shows the brighter parts of the nebulosity as described.

Lord Rosse (*Phil. Trans.*, 1861, *p.* 733) gives a brief description of the nebula.

The Photograph shows that the parts described by Sir J. Herschel form only a small portion of a gigantic nebula of an irregular oval character. It is more than two degrees in length from *north* to *south*, and forty-seven minutes of arc in breadth on the *following* side of 52 (*k*) Cygni, which it barely touches by a slight projection of the nebulosity.

The streaky character of the nebulosity will be observed on the print, and when an opportunity occurs to take another photograph with a much longer exposure of the plate, the connections of the several parts of the nebulosity will doubtless be shown.

The bright side of the nebula seems to form a sharply-defined boundary between the stream of the Milky Way stars and those on its *preceding* side.

The straight line crossing the star 52 *Cygni* was caused by a sudden gust of wind during the exposure of the plate.

Plate 46.

NEBULA H I 192 CEPHEI.

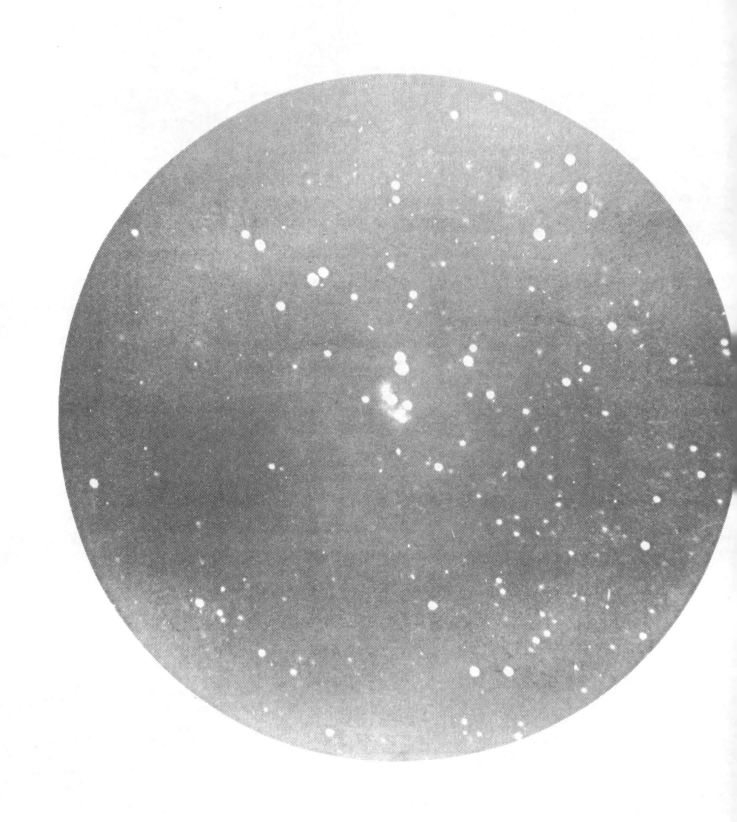

Nebula H̤. I. 192 Cephei.

Diameter of the field—19 minutes of arc.

Scale—1 millimetre to 6 seconds of arc.

Co-ordinates of the centre of the nebula for the epoch A.D. 1900.

R.A. 20h. 57m. 38·8s. ; Dec. N. 54° 8·5′ (= D.M. 2533—Zone+53°, mag. 8·7).

The photograph was taken with the 20-inch reflector on August 21st, 1892, between sidereal time 21h. 21m. and 22h. 51m., with an exposure of the plate during ninety minutes.

REFERENCES.

N.G.C. 7008.   G.C. 4627.   *h* 2099.

Sir J. Herschel, in the G.C., describes the nebula as considerably bright; large; extended 45° ± ; barely resolvable; double star attached.

Lord Rosse (*Phil. Trans.*, 1861, *p.* 734, *pl. XXX., fig.* 37) describes and delineates the nebula much as it is shown on the photograph, but the positions of the stars do not agree.   A further reference is also made to it in *Obs. of Neb. and Cl.*, *p.* 159.

The Photograph shows the nebula and the stars as they appeared on the date given above, with an exposure of ninety minutes; and another photograph which was taken on October 8th, 1893, with an exposure of three hours, shows an extension of the nebulosity on the *south following* side.

Plate 47.

CLUSTER MESSIER 15 PEGASI.

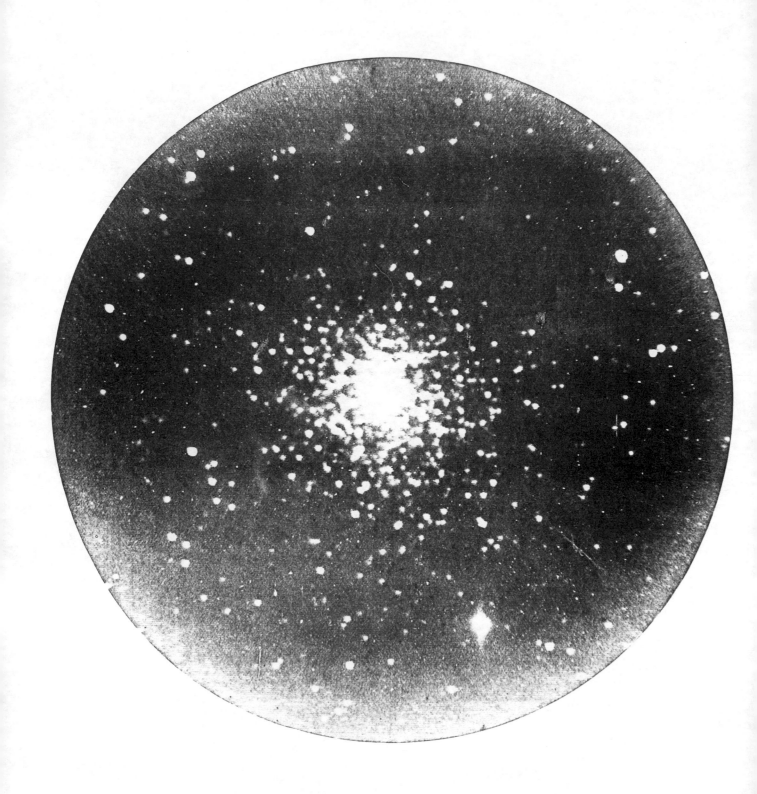

Cluster Messier 15 Pegasi.

R.A. 21h. 25m. 9s.; Dec. N. 11° 43·7′.

Diameter of the field—19 minutes of arc.

Scale—1 millimetre to 6 seconds of arc.

Co-ordinates of the Fiducial stars marked with dots for the epoch A.D. 1900.

Star (.) D.M. No. 4578—Zone+11° ... R.A. 21h. 25m. 20·4s. ... Dec. N. 11° 50·4′ ... Mag. 7·8
,, (··) ,, ,, 4579 ,, 11° ... ,, 21h. 25m. 33·7s. ... ,, 11° 40·3′ ... ,, 9·5

The photograph was taken with the 20-inch reflector on November 4th, 1890, between sidereal time 22h. 2m. and 24h. 2m., with an exposure of the plate during two hours.

REFERENCES.

N.G.C. 7078. G.C. 4670. h 2120.

Sir J. Herschel, in the G.C., p. 130, describes it as a remarkable globular cluster; very bright; very large; irregularly round; very suddenly much brighter in the middle; well resolved; stars very small.

Lord Rosse (*Obs. of Neb. and Cl., p. 161*) describes it as a globular cluster with bright and faint stars sharply distinguished from each other. Nucleus, a little eccentric, forms the middle.

The Photograph confirms the general descriptions given above, and the negative shows, separately, the stars of which the cluster is composed distinctly through the nebulosity to the centre. The nebulosity is too dense to print through, and measures about 200 seconds of arc in diameter. Many of the stars have a nebulous appearance, and they are arranged in curves, lines, and patterns of various forms, with lanes or spaces between them.

Plate 48.

REGION OF NOVA CYGNI.

REGION OF NOVA CYGNI.

Region of Nova Cygni.

R.A. 21h. 37m. 47s. ; Dec. N. 42° 23·9′.

The photograph covers the region between R.A. 21h. 34m. 20s. and R.A. 21h. 41m. 39s. Declination between 41° 33′ and 43° 14′ North.

Scale—1 millimetre to 24 seconds of arc.

Co-ordinates of the Fiducial stars marked with dots for the epoch A.D. 1900.

Star (.) D.M. No. 4223—Zone +41°	...	R.A. 21h. 35m. 19·7s.	...	Dec. N. 42° 6·0′	...	Mag. 9·0
,, (··) ,, ,, 4177 ,, 42°	...	,, 21h. 36m. 15·2s.	...	,, 42° 49·3′	...	,, 5·2
,, (∵) ,, ,, 4243 ,, 41°	...	,, 21h. 38m. 36·0s.	...	,, 42° 59·0′	...	,, 7·0
,, (∷) ,, ,, 4188 ,, 42°	...	,, 21h. 38m. 38·4s.	...	,, 42° 32·5′	...	,, 8·0

The photograph was taken with the 20-inch reflector on September 27th, 1891, between sidereal time 21h. 47m. and 23h. 47m., with an exposure of the plate during two hours.

Nova Cygni was discovered by Dr. J. F. J. Schmidt at Athens, on November 24th, 1876, and its co-ordinates for 1878 are given, R.A. 21h. 36m. 55s. ; Dec. North, 42° 17·9′.

Drs. Copeland and Lohse published in *Copernicus, Vol. II.*, a Memoir and Chart of the region of the *Nova*, and the results of a comparison made between the chart and the photograph are recorded in the *Monthly Notices, R.A.S., Vol. LII., pp.* 372, 373. There is a statement on p. 373, that " the *Nova* is not given on the chart," the meaning of which is that it is not given in the same form as the other stars, but in the form of a diagram. Details are given in the table which is included in the Memoir (p. 114) of the magnitude of the *Nova* from January 8th, 1877, when it was 6·7, to March 24th, 1882, when it was of the 14th magnitude.

On the Photograph, the *Nova* appears as a star of about 13th magnitude, and I am indebted to Drs. Copeland and Lohse for pointing out to me its place among the stars on the photograph, which is at the centre of the circle drawn round it.

Plate 49.

NEBULA ♄ I 53 PEGASI.

Nebula ♄. I. 53 Pegasi.

R.A. 22h. 32m. 29s. ;   Dec. N. 33° 53·9′.

The photograph covers the region between R.A. 22h. 29m. 12s. and R.A. 22h. 35m. 37s. Declination between 33° 3′ and 34° 44′ North.

Scale—1 millimetre to 24 seconds of arc.

Co-ordinates of the Fiducial stars marked with dots for the epoch A.D. 1900.

Star (.) D.M. No. 4542—Zone+ 33°	...	R.A. 22h. 30m. 52·5s.	...	Dec. N. 33° 55·9′	...	Mag. 9·0
„ (··) „ „ 4545 „ 33°	...	„ 22h. 31m. 30·3s.	...	„ 34° 6·5′	...	„ 9·1
„ (∵) „ „ 4547 „ 33°	...	„ 22h. 31m. 46·5s.	...	„ 33° 34·3′	...	„ 9·0
„ (∷) „ „ 4731 „ 34°	...	„ 22h. 32m. 38·8s.	...	„ 34° 24·2′	...	„ 8·4

The photograph was taken with the 20-inch reflector on October 18th, 1892, between sidereal time 22h. 18m. and 2h. 5m., with an exposure of the plate during three hours and thirty-nine minutes.

REFERENCES.

N.G.C. 7331. G.C. 4815. _h_ 2172.

Sir J. Herschel, in the G.C., p. 132, describes the nebula as bright ; pretty large ; pretty much extended, 160° ; gradually much brighter in the middle.

Lord Rosse (_Phil. Trans._, 1861, _p._ 734, _pl._ XXX., _fig._ 39) gives a drawing of the nebula, and in the _Obs. of Neb. and Cl._, _pp._ 166, 167, refers to it in considerable detail, but with doubts as to the reality of some of the observations.

The Photograph shows the nebula to be symmetrically oval, with a stellar nucleus surrounded by dense nebulosity, and round the nebulosity faint rings of nebulous matter somewhat resembling, on a small scale, the nebula in _Andromeda_, but a long exposure of the plate will be required to show these satisfactorily.

Plate 50.

NEBULA ♄ I 55 PEGASI.

NEBULA ♄ I 55 PEGASI.

Nebula ♄. I. 55 Pegasi.

R.A. 22h. 59m. 56s. ; Dec. N. 11° 47·0′.

The photograph covers the region between R.A. 22h. 57m. 8s. and R.A. 23h. 2m. 40s. Declination between 10° 59′ and 12° 40′ North.

Scale—1 millimetre to 24 seconds of arc.

Co-ordinates of the Fiducial stars marked with dots for the epoch A.D. 1900.

Star (.) D.M, No. 4926—Zone+11°	...	R.A. 22h. 58m. 44·7s.	...	Dec. N. 11° 20·2′	...	Mag. 8·2
,, (··) ,, ,, 4932 ,, 12°	...	,, 22h. 59m. 11·6s.	...	,, 12° 17·3′	...	,, 8·6
,, (·˙) ,, ,, 4929 ,, 11°	...	,, 22h. 59m. 39·0s.	...	,, 11° 38·0′	...	,, 9·5
,, (::) ,, ,, 4938 ,, 11°	...	,, 23h. 2m. 27·9s.	...	,, 11° 15·7′	...	,, 8·6

The photograph was taken with the 20-inch reflector on October 22nd, 1892, between sidereal time 20h. 51m. and 2h. 3m., with an exposure of the plate during four hours.

REFERENCES.

N.G.C. 7479.   G.C. 4892.   h 2205.

Sir J: Herschel (*Phil. Trans.*, 1833, *p.* 475, *pl. XIV.*, *fig.* 39) describes the nebula as pretty bright ; irregularly round ; resolved ; two or three stars in it ; extended between two stars. The drawing does not resemble the photograph.

Lord Rosse (*Phil. Trans.*, 1850, *p.* 511, *pl. XXXVI.*, *fig.* 4) is uncertain whether it should be classed as a spiral or as an annular nebula. The drawing gives a fair idea of part of its form. In the *Obs. of Neb. and Cl.*, *p.* 170, several observations of the nebula are recorded confirming the accuracy of the drawing referred to above.

The Photograph shows the nebula to be elliptical, with a dense, broad line of nebulosity, curved at both ends, forming the major axis, which has a star of about 15th magnitude in its centre, and there is also a slightly fainter star in the centre of the *preceding* semi-ellipse. No structure is visible in it, such as that shown on the drawing by Lord Rosse, but the semi-ellipse on the *following* side is shown on the photograph, though it is not on the drawing.

Plate 51.

CLUSTER MESSIER 52 CEPHEI.

Cluster Messier 52 Cephei.

R.A. 23h. 19m. 49s.; Dec. N. 61° 2·9′.

The photograph covers the region between R.A. 23h. 14m. 53s. and R.A. 23h. 25m. 23s. Declination between 60° 11′ and 61° 52′ North.

Scale—1 millimetre to 24 seconds of arc.

Co-ordinates of the Fiducial stars marked with dots for the epoch A.D. 1900.

Star (.) D.M. No. 2521—Zone + 60°	...	R.A. 23h. 15m. 36·9s.	...	Dec. N. 60° 36·0′	...	Mag. 7·0
,, (··) ,, ,, 2444 ,, 61°	...	,, 23h. 20m. 26·0s.	...	,, 61° 43·4′	...	,, 5·3
,, (∵) ,, ,, 2540 ,, 60°	...	,, 23h. 21m. 51·0s.	...	,, 60° 32·0′	...	,, 7·0
,, (::) ,, ,, 2454 ,, 61°	...	,, 23h. 23m. 35·0s.	...	,, 61° 22·2′	...	,, 7·9

The photograph was taken with the 20-inch reflector on August 31st, 1892, between sidereal time 23h. 10m. and 24h. 10m., with an exposure of the plate during one hour.

REFERENCES.

N.G.C. 7654. G.C. 4957. *h* 2238.

Sir J. Herschel, in the G.C., p. 135, describes the cluster as large; rich; much compressed in the middle; round; stars 9th to 13th magnitude.

Lord Rosse (*Obs. of Neb. and Cl., p.* 173) describes it as a splendid cluster; pretty compressed; field uniformly dotted with stars; about 200 in number.

The Photograph fully bears out the descriptions given, and shows each star in accurate relative position and magnitude.

Plate 52.

CLUSTER ♄ VI 30 CASSIOPEIÆ.

Cluster ♅. VI. 30 Cassiopeiæ.

R.A. 23h. 51m. 59s. ; Dec. N. 56° 9·6′.

The photograph covers the region between R.A. 23h. 47m. 16s. and R.A. 23h. 56m. 59s. Declination between 55° 22′ and 57° 3′. North.

Scale—1 millimetre to 24 seconds of arc.

Co-ordinates of the Fiducial stars marked with dots for the epoch A.D. 1900.

Star (.) D.M. No. 3106—Zone+56°	...	R.A. 23h. 48m. 39·1s.	...	Dec. N. 56° 16·2′	...	Mag. 8·0
„ (··) „ „ 3045 „ 55°	...	„ 23h. 50m. 23·6s.	...	„ 55° 52·9′	...	„ 8·9
„ (∵) „ „ 3119 „ 56°	...	„ 23h. 52m. 44·0s.	...	„ 56° 34·4′	...	„ 7·7
„ (∷) „ „ 3129 „ 56°	...	„ 23h. 55m. 13·2s.	...	„ 56° 17·8′	...	„ 8·3

The photograph was taken with the 20-inch reflector on November 26th, 1892, between sidereal time 0h. 39m. and 2h. 9m., with an exposure of the plate during ninety minutes.

REFERENCES.

N.G.C. 7789.   G.C. 5031.   h 2284.

Sir J. Herschel (*Phil. Trans.*, 1833, *p.* 480) describes it as a most superb cluster, which fills the field and is full of stars; gradually brighter in the middle, but no condensation to a nucleus. Stars 11th to 18th magnitude. In the G.C., p. 137, is also a reference to the cluster.

Lord Rosse (*Obs. of Neb. and Cl., p.* 176) describes it as a very coarse cluster, stars quite distinct and no nebulosity visible, dark holes and jagged branches, but no regular arrangement.

The Photograph confirms the descriptions just cited, and in addition shows patterns, consisting of lines, wreaths, and curves of stars, each one charted in its true relative position and magnitude in the cluster as well as over the whole region. The co-ordinates of the position of any star among the thousands shown on the chart can readily be determined, within a few seconds of arc, by measurements from the Fiducial stars.

Plate 53.

NEBULA ♆ II 240 PEGASI.

Nebula ♄. II. 240 Pegasi.

R.A. 23h. 58m. 8s.; Dec. N. 15° 34·5′.

The photograph covers the region between R.A. 23h. 56m. 22s. and R.A. 23h. 59m. 58s. Declination between 15° 4′ and 16° 7′ North.

Scale—1 millimetre to 15 seconds of arc.

Co-ordinates of the Fiducial stars marked with dots for the epoch A.D. 1900.

Star (.) D.M. No. 4924—Zone +15°	...	R.A. 23h. 57m. 16·2s.	...	Dec. N. 15° 53·1′	...	Mag. 9·1
„ (··) „ „ 4927 „ 15°	...	„ 23h. 57m. 34·2s.	...	„ 15° 28·7′		„ 9·0
„ (∴) „ „ 4933 „ 15°	...	„ 23h. 58m. 59·1s.	...	„ 15° 33·9′	...	„ 9·5
„ (∷) „ „ 5092 „ 14°	...	„ 23h. 59m. 20·1s.	...	„ 15° 12·5′	...	„ 8·8

The photograph was taken with the 20-inch reflector on December 2nd, 1891, between sidereal time 0h. 9m. and 2h. 50m., with an exposure of the plate during two hours and thirty-six minutes.

REFERENCES.

N.G.C. 7814. G.C. 5046. *h* 2297.

Sir J. Herschel, in the G.C., describes the nebula as considerably bright; considerably large; very gradually brighter in the middle.

Lord Rosse (*Phil. Trans.*, 1861, *p.* 736, *pl. XXX., fig.* 42) states that a decided dark lane runs through it in the direction of its major axis. The drawing gives a very good representation of the nebula and agrees with the photograph, excepting in the widening out of the extremities of the nebulosity. In the *Obs. of Neb. and Cl., p.* 177, *pl. V., fig.* 11, further references are made, and another drawing given, but they do not add materially to those referred to in the foregoing paragraph.

The Photograph very fully confirms the observations and the drawings made by Lord Rosse, excepting, as before said, the widening out of the faint nebulosity at each extremity of the major axis, which the photograph does not show. The dark lane is conspicuously visible, and it divides the major axis into two equal parts. The central part of the nebula seems to be a globular mass, and measures 41 seconds of arc in diameter, but it differs from a star image in not having a central condensation or nucleus. The dark lane measures about 8 seconds of arc in breadth, which is uniformly maintained as far as it can be traced on the photograph, except at the place where it crosses the denser part of the globular mass. There it is narrower and less distinctly visible.

The inferences that I draw from the photograph are, that the dark lane is a ring, seen edgewise, surrounding the nebula, and that there is within it another very wide, nebulous, faint ring surrounding the central mass. The outer dark ring intercepts the light of the inner nebulous ring, and where it crosses the luminous central condensation the light is intercepted in a greater degree.

DEDUCTIONS FROM THE PHOTOGRAPHS.

THE references which have been made to the photographs in the preceding pages of this work contain very little that is of an inferential or of a speculative character, though it is obvious that there is considerable material available for the theorist. I may therefore invite special attention to the nebulæ and clusters of stars referred to in the following paragraphs :—

1st. If we examine the spiral nebulæ, ⩍. I. 56-57 *Leonis, pl.* 25; *M.* 81 *Ursæ Majoris, pl.* 26; ⩍. I. 168 *Ursæ Majoris, pl.* 27; *M.* 51 *Canum Venaticorum pl.* 30; and *M.* 101 *Ursæ Majoris, pl.* 32, we shall see in each of them (clearly on the negatives) a stellar nucleus surrounded by more or less dense nebulosity; and with each nucleus as a centre, the convolutions of the spirals are symmetrically formed around it. We also see that the convolutions of the spirals are broken up into well-defined stars, or into star-like condensations, and that there is residual dense nebulosity in the convolutions, and faint nebulosity between them.

These facts lead us to consider the question—To what cause can these uniform effects be referred ?

We may not, with the limited amount of evidence before us, be very positive, but the following suggestions may be worthy of consideration :—

(*a.*) If two masses of matter of a gaseous or of a meteoric character should collide in space, would not the effect be the production of a vortical motion, and the result be a spiral ?

(*b.*) If two globular clusters meet and collide, would not the result also be a spiral, though the conditions in this case are the converse of those suggested in the first question; for in this would be involved the partial or total breaking up of some of the stars in the cluster and their conversion into gaseous or into finely divided matter ?

2nd. If we examine the globular clusters, *M.* 53 *Comæ Berenicis, pl.* 29; *M.* 3 *Canum Venaticorum, pl.* 31; *M.* 5 *Libræ, pl.* 33; *M.* 13 *Herculis, pl.* 34; *M.* 12 *Ophiuchi, pl.* 35; *M.* 10 *Ophiuchi, pl.* 36; *M.* 92 *Herculis, pl.* 37; *M.* 15 *Pegasi, pl.* 47, we shall see in these clusters, as they appear on the photographs, and further as studied by aid of the negatives, various patterns, lines, and complicated curves of stars, so arranged that with very little stress on the imagination we perceive that some of the general characteristics of the spiral nebulæ are traceable in them; while the whole of the nebulosity has not been absorbed or utilized in the formation of the stars. Some of the nebulosity, varying in extent and density, is shown on the photographs to be present in each cluster.

Is it then probable that these clusters have had their origin in the condensation of Spiral Nebulæ?

3rd. There are clusters of stars that suggest a further development of the globular clusters, and if we refer to *M*. 11 *Antinoi, pl.* 40; *M*. 37 *Aurigæ, pl.* 18; and to the two clusters *H. VI*. 33 *and* 34 *Persei, pl.* 7, we shall see in each of them appearances strongly resembling those in the globular clusters, but without the nebulosity. There is not a trace of nebulosity shown in one of them even after an exposure of the plate during three hours in the 20-inch reflector.

4th. The formation of globular clusters from spiral nebulæ is not the only process that is indicated by the photographs; for if we turn to the *Dumb-bell Nebula* in *Vulpecula, pl.* 44, we see the globular mass of nebulosity, of which it is composed, apparently condensing into a globular cluster of stars, many of them already being partly formed.

Whatever value may be attached to these suggestions, and whether they are true or the reverse, in either case they do not affect the unquestionable accuracy of the photographic evidence which forms the substance of this publication.

Printed in the United States
By Bookmasters